T0140291

Smart Innovation, Systems and Technologies

Volume 24

Series Editors

R. J. Howlett, Shoreham-by-Sea, UK
L. C. Jain, Adelaide, Australia

For further volumes:
http://www.springer.com/series/8767

George A. Tsihrintzis · Maria Virvou
Lakhmi C. Jain
Editors

Multimedia Services
in Intelligent Environments

Advances in Recommender Systems

 Springer

Editors
George A. Tsihrintzis
Maria Virvou
Department of Informatics
University of Piraeus
Piraeus
Greece

Lakhmi C. Jain
School of Electrical and Information
  Engineering
University of South Australia
Adelaide, SA
Australia

ISSN 2190-3018   ISSN 2190-3026 (electronic)
ISBN 978-3-319-03340-2   ISBN 978-3-319-00372-6 (eBook)
DOI 10.1007/978-3-319-00372-6
Springer Cham Heidelberg New York Dordrecht London

Printed on acid-free paper

Springer is part of Springer Science+Business Media (www.springer.com)

Foreword

I have always believed that the most fertile ground for innovation is the "valley" between two or more active research areas. This belief has shaped my research activity so far, guiding me in exciting journeys between machine learning and personalization, among others. The editors of this book, and its three predecessors in the series, share the same enthusiasm for cross-fertilization of different research domains. Having themselves contributed significantly in different areas, ranging from pattern recognition to user modeling and intelligent systems, they decided to present different views upon *Multimedia Systems and Services, including tools and methodologies, software development, system integration and now recommender systems.*

Recent advances in Information and Communication Technologies (ICT) have increased the computational power of computers, while embedding them in various mobile devices, thus increasing enormously the possibilities of communication and transfer of data. These developments have created new opportunities for generating, storing and sharing multimedia, such as audio, image, video, animation, graphics, and text. YouTube for example, has more than 1 billion unique visitors each month, uploading 72 hours of video every minute! Providing the infrastructure to make this possible was clearly the first step, but at this volume of data one needs to urgently address the issue of retrieving the most suitable information for each user efficiently. Failure to do so may result in disappointment of the users in the near future and a major set-back in this revolutionary social development.

Personalization is among the most convincing proposals in addressing this problem of information overload. In particular recommender systems provide simple and very effective ways of understanding the needs and interests of individuals and selecting the content they would be more interested in. Such systems have been used to assist users in selecting movies, music, books and content in various other forms. The blending of the two very different research areas, multimedia systems and recommender systems, seems to be providing a new very fertile field for innovation.

In this book, the editors and the authors of the individual chapters present recent advances in designing and applying recommender systems to multimedia systems and services. I consider this particularly timely, as the need to address the overload with multimedia information has sparked research activity throughout the world. Therefore, the book is expected to provide a good overview of current research to newcomers in the field and help them gain insights on how to add value to multimedia systems and services through intelligent content recommendation. An exciting research and innovation arena that is still being shaped.

Georgios Paliouras

Research Director and Head of the Division

of Intelligent Information Systems

National Centre for Scientific Research "Demokritos"

Institute of Informatics and Telecommunications

Athens, Greece

Preface

Multimedia services are now commonly used in various activities in the daily lives of humans. Related application areas include services that allow access to large depositories of information, digital libraries, e-learning and e-education, e-government and e-governance, e-commerce and e-auctions, e-entertainment, e-health and e-medicine, and e-legal services, as well as their mobile counterparts (i.e., *m-services*). Despite the tremendous growth of multimedia services over the recent years, there is an increasing demand for their further development. This demand is driven by the ever-increasing desire of society for easy accessibility to information in friendly, personalized, and adaptive environments. With this view in mind, we have been editing a series of books on *Multimedia Services in Intelligent Environments* [1–9].

Specifically, this book is a continuation of our previous books [1–3]. In this book, we examine recent *Advances in Recommender Systems*. Recommender systems are crucial in multimedia services, as they aim at protecting the service users from *information overload*. The book includes nine chapters, which present various recent research results in recommender systems. Each chapter in the book was reviewed by two independent reviewers for novelty and clarity of the research presented in it. The reviewers' comments were incorporated in the final version of the chapters.

This research book is directed to professors, researchers, application engineers, and students of all disciplines. We hope that they all will find it useful in their works and researches.

We are grateful to the authors and the reviewers for their excellent contributions and visionary ideas. We are also thankful to Springer for agreeing to publish this book. Last, but not least, we are grateful to the Springer staff for their excellent work.

Piraeus, Greece	George A. Tsihrintzis
Piraeus, Greece	Maria Virvou
Adelaide, Australia	Lakhmi C. Jain

References

1. Tsihrintzis, G.A., Jain, L.C. (eds.): Multimedia Services in Intelligent Environments—Advanced Tools and Methodologies. Studies in Computational Intelligence Book Series, vol. 120. Springer, (2008)
2. Tsihrintzis, G.A., Virvou, M., Jain, L.C. (eds.): Multimedia Services in Intelligent Environments—Software Development Challenges and Solutions. Smart Innovation, Systems, and Technologies Book Series, vol. 2. Springer, (2010)
3. Tsihrintzis, G.A., Jain, L.C. (eds.): Multimedia Services in Intelligent Environments—Integrated Systems. Smart Innovation, Systems, and Technologies Book Series, vol. 3. Springer, (2010)
4. Tsihrintzis, G.A., Virvou, M., Howlett, R.J., Jain, L.C. (eds.): New Directions in Intelligent Interactive Multimedia Systems and Services. Studies in Computational Intelligence Book Series, vol. 142. Springer, (2008)
5. Tsihrintzis, G.A., Damiani, E., Virvou, M., Howlett, R.J., Jain, L.C. (eds.): Intelligent Interactive Multimedia Systems and Services. Smart Innovation, Systems, and Technologies Book Series, vol. 6. Springer, (2010)
6. Tsihrintzis, G.A., Virvou, M., Jain, L.C., Howlett, R.J. (eds.): Intelligent Interactive Multimedia Systems and Services. Smart Innovation, Systems, and Technologies Book Series, vol. 11. Springer, (2011)
7. Virvou, M., Jain, L.C. (eds.): Intelligent Interactive Systems in Knowledge-Based Environments. Studies in Computational Intelligence Book Series, vol. 104. Springer, (2008)
8. Obaidat, M., Tsihrintzis, G.A., Filipe, J. (eds.): e-Business and Telecommunications: 7th International Joint Conference, ICETE, Athens, Greece, July 26–28, 2010 Revised Selected Papers. Communications in Computer and Information Science Book Series, vol. 222. Springer, (2012)
9. Tsihrintzis, G.A., Pan, J.-S., Huang, H.C., Virvou, M., Jain, L.C. (eds.): Proceedings of the Eighth International Conference on Intelligent Information Hiding and Multimedia Signal Processing (IIH-MSP 2012). IEEE Conference Publishing Services, (2012)

Contents

Multimedia Services in Intelligent Environments: Advances in Recommender Systems

George A. Tsihrintzis, Maria Virvou and Lakhmi C. Jain

Abstract Multimedia services are now commonly used in various activities in the daily lives of humans. Related application areas include services that allow access to large depositories of information, digital libraries, e-learning and e-education, egovernment and e-governance, e-commerce and e-auctions, e-entertainment, e-health and emedicine, and e-legal services, as well as their mobile counterparts (i.e., m-services). Despite the tremendous growth of multimedia services over the recent years, there is an increasing demand for their further development. This demand is driven by the everincreasing desire of society for easy accessibility to information in friendly, personalized and adaptive environments. With this view in mind, we have been editing a series of books on Multimedia Services in Intelligent Environments [1-4]. This book is the fourth in this series. In this book, we examine recent Advances in Recommender Systems, which are crucial in multimedia services, as they aim at protecting the service users from information overload.

1 Introduction

The term *Multimedia Services* has been coined to refer to services which make use of coordinated and secure storage, processing, transmission, and retrieval of information which exists in various forms. As such, the term refers to several levels of data processing and includes such diverse application areas as digital libraries, e-learning, e-government, e-commerce, e-entertainment, e-health, and e-legal services, as well as their mobile counterparts (i.e., *m-services)*. As new

G. A. Tsihrintzis (✉) · M. Virvou
University of Piraeus, Piraeus, Greece
e-mail: geoatsi@unipi.gr

L. C. Jain
University of South Australia, Adelaide, Australia

G. A. Tsihrintzis et al. (eds.), *Multimedia Services in Intelligent Environments*,
Smart Innovation, Systems and Technologies 24, DOI: 10.1007/978-3-319-00372-6_1,
© Springer International Publishing Switzerland 2013

multimedia services appear constantly, new challenges for advanced processing arise daily. Thus, we have been attempting to follow relevant advances in a series of edited books on multimedia services in intelligent environments. This is the fourth volume on the topic. In our earlier books [1–3], we covered various aspects of processing in multimedia services.

More specifically, in [1] we were concerned mostly with low level data processing in multimedia services in intelligent environments, including storage (Chap. 2), recognition and classification (Chaps. 3 and 4), transmission (Chaps. 5 and 6), retrieval (Chaps. 7 and 8), and securing (Chaps. 9 and 10) of information. Four additional chapters in [1] presented intermediate level multimedia services in noise and hearing monitoring and measuring (Chap. 11), augmented reality (Chap. 12), automated lecture rooms (Chap. 13) and rights management and licensing (Chap. 14). Finally, Chap. 15 was devoted to a high-level intelligent recommender service in scientific digital libraries.

In [2], we were concerned with various software development challenges and related solutions that are faced when attempting to accommodate multimedia services in intelligent environments. Specifically, [2] included an editorial introduction and ten additional chapters, as follows: Chap. 2 by Savvopoulos and Virvou was on "Evaluating the generality of a life-cycle framework for incorporating clustering algorithms in adaptive systems." Chapter 3 by Chatterjee, Sadjadi and Shu-Ching Chen dealt with "A Distributed Multimedia Data Management over the Grid." Chapter 4 was authored by Pirzadeh and Hamou-Lhadj and covered "A View of Monitoring and Tracing Techniques and their Application to Service-based Environments." Chapter 5 by Bucci, Sandrucci, and Vicario was devoted to "An ontological SW architecture supporting the contribution and retrieval of Service and Process models." Chapter 6 by D'Ambrogio dealt with "Model-driven Quality Engineering of Service-based Systems." Chapter 7 by Katsiri, Serrano, and Serat dealt with "Application of Logic Models for Pervasive Computing Environments and Context-Aware Services Support," while Chap. 8 by Patsakis and Alexandris covered "Intelligent Host Discovery from Malicious Software."

In [2], we also included three chapters on new theoretical results, development methodologies and tools which hold promise to be useful in the development of future systems supporting multimedia services in intelligent environments. Specifically, Chap. 9 by Fountas dealt with "Swarm Intelligence: The Ant Paradigm," while Chap. 10 by Artikis dealt with "Formulating Discrete Geometric Random Sums for Facilitating Intelligent Behaviour of a Complex System under a Condition of Major Risk." Finally, Chap. 11 by Artikis dealt with "Incorporating a Discrete Renewal Random Sum in Computational Intelligence and Cindynics."

In [3], we presented various integrated systems that were developed to accommodate multimedia services in intelligent environments. Besides the editorial introduction, [3] included thirteen additional chapters, as follows: Chaps. 2 and 3 were devoted to multimedia geographical information systems. Specifically, Chap. 2 by Gemizi, Tsihrintzis, and Petalas was on "Use of GIS and Multi-Criteria

Evaluation Techniques in Environmental Problems," while Chap. 3 by Charou, Kabassi, Martinis, and Stefouli was on "Integrating Multimedia GIS Technologies in a Recommendation System for Geotourism." Chapters 4 and 5 covered aspects of e-entertainment systems. Specifically, Chap. 4 by El-Nasr and Zupko was on "Lighting Design Tools for Interactive Entertainment," while Chap. 5 by Szczuko and Kostek was on "Utilization of Fuzzy Rules in Computer Character Anima-tion". Chapters 6, 7 covered aspects of education and e-learning systems. Spe-cifically, Chap. 6 by Nakatani, Tsumaki, and Tamai was on "Instructional Design of a Requirements Engineering Education for Professional Engineers," while Chap. 7 by Burdescu and Mihăescu was on "Building Intelligent e-Learning Systems by Activity Monitoring and Analysis." Chapters 8, 9 were devoted to medical diagnosis systems. Specifically, Chap. 8 by Schmidt and Vorobieva was on "Supporting the Search for Explanations of Medical Exceptions," while Chap. 9 by Aupet, Garcia, Guyennet, Lapayre, and Martins was on "Security in Medical Telediagnosis."

Chapters 10, 11 were devoted to telemonitoring systems. Specifically, Chap. 10 by Żwan, Sobala, Szczuko, and Czyzewski was on "Audio Content Analysis in the Urban Area Telemonitoring System," while Chap. 11 by Dalka, Szwoch, Szczuko, and Czyzewski was on "Video Content Analysis in the Urban Area Telemonitoring System." Chapter. 12 by Karapiperis, Stojanovic, Anicic, Apostolou and Despotis, was on "Enterprise Attention Management." An additional chapter, namely Chap. 13 by Raij and Lehto, was on "e-Welfare as a Client-driven Service Con-cept." Finally, Chap. 14 by Panoulas, Hadjileondiadis, and Panas was on "Brain-Computer Interface (BCI): Types, Processing Perspectives and Applications."

The book at hand is a continuation of our coverage of multimedia services in intelligent environments. In particular, this book is devoted to a specific class of software systems, called *Recommender Systems*, which are analyzed in the fol-lowing section.

2 Recommender Systems

Recent advances in electronic media and computer networks have allowed the creation of large and distributed repositories of information. However, the immediate availability of extensive resources for use by broad classes of computer users gives rise to new challenges in everyday life. These challenges arise from the fact that users cannot exploit available resources effectively when the amount of information requires prohibitively long user time spent on acquaintance with and comprehension of the information content. Thus, the risk of information overload of users imposes new requirements on the software systems that handle the information. Such systems are called *Recommender Systems* and attempt to pro-vide information in a way that will be most appropriate and valuable to its users and prevent them from being overwhelmed by huge amounts of information that, in the absence of recommender systems, they should browse or examine [4].

Besides the current editorial chapter, the book includes an additional eight (8) chapters organized into two parts. The first part of the book consists of Chaps. 2, 3, 4 and 5. In this part, we address various aspects of recommender systems which form the core of recommendation services. Specifically, Chap. 2 by Lampropoulos and Tsihrintzis is on "A Survey of Approaches to Designing Recommender Systems." Chapter 3 by Nguyen and Santos is on "Hybrid User Model For Capturing a User's Information Seeking Intent." Chapter 4 by Lamb, Randles and Al-Jumeily is on "Recommender Systems: Network Approaches." Chapter 5 by Felfernig, Jeran, Ninaus, Reinfrank, and Reiterer makes contributions "Towards the Next Generation of Recommender Systems: Applications and Research Challenges."

The second part of the book consists of four chapters and presents various new theoretical results and tools that are expected to be incorporated in and improve recommender systems and recommendation services. Chapter 6 by Toledo, Sookhanaphibarn, Thanwonmas, and Rinaldo is on "Content-based Recommendation for Stacked-Graph Navigation," while Chap. 7 by Sakamoto and Kuboyama is on "Pattern Extraction from Graphs and Beyond." Chapter 8 by Kinoshita is on "Dominant Adaptive Hierarchical Process as Measuring Method of Service Values," while, finally, Chap. 9 by Artikis and Artikis is on "Applications of a Stochastic Model in Supporting Intelligent Multimedia Systems and Educational Processes."

3 Conclusions

To avoid the risk of information overload of users, new requirements are imposed on the software systems that handle the information. Such systems are called *Recommender Systems* and attempt to provide information in a way that will be most appropriate and valuable to its users and prevent them from being overwhelmed by huge amounts of information that, in the absence of recommender systems, they should browse or examine. In this book, we have investigated recommender systems and attempted to both survey the field of methodologies incorporated in current recommender systems and look at future recommender systems and new, more advanced recommendation methodologies.

As the field of recommendation cannot be covered in one volume, we have devoted our next book in this series to *Recommendation Services* [4]. In there, we examine a variety of specific recommendation applications, from fields such as biology and medicine, education, mobile computing, cultural heritage, tourism, etc.

References

1. Tsihrintzis, G.A., Jain, L.C. (eds.): Multimedia Services in Intelligent Environments— Advanced Tools and Methodologies, Studies in Computational Intelligence Book Series. Vol. 120. Springer, Berlin-Heidelberg (2008)

2. Tsihrintzis, G.A., Virvou, M., Jain, L.C. (eds.): Multimedia Services in Intelligent Environments—Software Development Challenges and Solutions, Smart Innovation, Systems, and Technologies Book Series. Vol. 2. Springer, Berlin-Heidelberg (2010)
3. Tsihrintzis, G.A., Jain, L.C. (eds.): Multimedia services in intelligent environments—integrated systems, Smart Innovation, Systems, and Technologies Book Series. Vol. 3. Springer, Berlin-Heidelberg (2010)
4. Tsihrintzis, G.A., Virvou, M., Jain, L.C. (eds.): Multimedia services in intelligent environments—Recommendation service. studies in computational intelligence book series. Springer, Berlin-Heidelberg (2013)

A Survey of Approaches to Designing Recommender Systems

Aristomenis S. Lampropoulos and George A. Tsihrintzis

Abstract The large amount of information resources that are available to users imposes new requirements on the software systems that handle the information. This chapter provides a survey of approaches to designing recommenders that address the problems caused by information overload.

1 Introduction to Recommender Systems

Recent advances in electronic media and computer networks have allowed the creation of large and distributed repositories of information. However, the immediate availability of extensive resources for use by broad classes of computer users gives rise to new challenges in everyday life. These challenges arise from the fact that users cannot exploit available resources effectively when the amount of information requires prohibitively long user time spent on acquaintance with and comprehension of the information content. Thus, the risk of information overload of users imposes new requirements on the software systems that handle the information. One of these requirements is the incorporation into the software systems of mechanisms that help their users when they face difficulties during human-computer interaction sessions or lack the knowledge to make decisions by themselves. Such mechanisms attempt to identify user information needs and to personalize human-computer interactions. (Personalized) Recommender Systems

A. S. Lampropoulos · G. A. Tsihrintzis (✉)
Department of Informatics, University of Piraeus,
80 Karaoli & Dimitriou, 18534 Piraeus, Greece
e-mail: geoatsi@unipi.gr

A. S. Lampropoulos
e-mail: arislamp@unipi.gr

G. A. Tsihrintzis et al. (eds.), *Multimedia Services in Intelligent Environments*,
Smart Innovation, Systems and Technologies 24, DOI: 10.1007/978-3-319-00372-6_2,
© Springer International Publishing Switzerland 2013

(RS) provide an example of software systems that attempt to address some of the problems caused by information overload.

RS are defined in [1] as software systems in which "people provide recommendations as inputs, which the system then aggregates and directs to appropriate recipients." Today, the term includes a wider spectrum of systems describing any system that provides individualization of the recommendation results and leads to a procedure that helps users in a personalized way to interesting or useful objects in a large space of possible options. RS form an important research area because of the abundance of their potential practical applications.

Clearly, the functionality of RS is similar to the social process of recommendation and reduction of information that is useless or uninteresting to the user. Thus, one might consider RS as similar to search engines or information retrieval systems. However, RS are to be differentiated from search engines or information retrieval systems as a RS not only finds results, but additionally uses its embedded individualization and personalization mechanisms to select objects (items) that satisfy the specific querying user needs. Thus, unlike search engines or information retrieval systems, a RS provides information in a way that will be most appropriate and valuable to its users and prevents them from being overwhelmed by huge amounts of information that, in the absence of RS, they should browse or examine. This is to be contrasted with the target of a search engine or an information retrieval system which is to "match" items to the user query. This means that a search engine or an information retrieval system tries to form and return *a ranked list* of all those items that match the query. Techniques of active learning such as *relevance-feedback* may give these systems the ability to refine their results according to the user preferences and, thus, provide a simple form of recommendation. More complex search engines such as *GOOGLE* utilize other kinds of criteria such as "*authoritativeness*", which aim at returning as many useful results as possible, but *not* in an individualized way.

A learning-based RS typically works as follows: (1) the recommender system collects all given recommendations at one place and (2) applies a learning algorithm, thereafter. Predictions are then made either with a model learnt from the dataset (model-based predictions) using, for example, a clustering algorithm [2, 3] or on the fly (memory-based predictions) using, for example, a nearest neighbor algorithm [2, 4]. A typical prediction can be a list of the top-N recommendations or a requested prediction for a single item [5].

Memory-based methods store training instances during training which are can be retrieved when making predictions. In contrast, model-based methods generalize into a model from the training instances during training and the model needs to be updated regularly. Then, the model is used to make predictions. Memory-based methods learn fast but make slow predictions, while model-based methods make fast predictions but learn slowly.

The roots of RS can be traced back to Malone et al. [6], who proposed three forms of filtering: cognitive filtering (now called content-based filtering), social filtering (now called collaborative filtering (CF)) and economic filtering. They also suggested that the best approach was probably to combine these approaches into the category of, so-called, *hybrid* RS.

1.1 Formulation of the Recommendation Problem

In general, the recommendation problem is defined as the problem of estimating ratings for the items that have not been seen by a user. This estimation is based on:

- ratings given by the user to other items,
- ratings given to an item by other users,
- and other user and item information (e.g. item characteristics, user demographics).

The recommendation problem can be formulated [7] as follows:

Let U be the *set of all users* $U = \{u_1, u_2, \ldots, u_m\}$ and let I be the *set of all possible items* $I = \{i_1, i_2, \ldots, i_n\}$ that can be recommended, such as music files, images, movies, etc. The space I of possible items can be very large.

Let f be a utility function that measures the usefulness of item i to user u,

$$f : U \times I \to R, \qquad (1)$$

where R is a totaly ordered set (e.g. the set of nonnegative integers or real numbers within a certain range). Then, for each user $u \in U$, we want to choose an item $i' \in I$ that maximizes the user utility function, i.e.

$$\forall u \in U, i'_u = \arg \max_{i \in I} f(u, i). \qquad (2)$$

In RS, the utility of an item is usually represented by a rating, which indicates how a particular user liked a particular item, e.g., user u_1 gave the object i_1 the rating of $R(1, 1) = 3$, where $R(u, i) \in \{1, 2, 3, 4, 5\}$.

Each user u_k, where $k = 1, 2, \ldots, m$, has a list of items I_{u_k} about which the user has expressed his/her preferences. It is important to note that $I_{u_k} \subseteq I$, while it is also possible for I_{u_k} to be the null set. This latter means that users are not required to express their preferences for all existing items.

Each element of the user space U can be defined with a profile that includes various user characteristics, such as age, gender, income, marital status, etc. In the simplest case, the profile can contain only a single (unique) element, such as User ID.

Recommendation algorithms enhance various techniques by operating

- either on *rows* of the matrix R, which correspond to ratings of a single user about different items,
- or on *columns* of the matrix R, which correspond to different users' ratings for a single item.

However, in general, the utility function can be an arbitrary function, including a profit function. Depending on the application, a utility f can either be specified by the user, as is often done for the user-defined ratings, or computed by the application, as can be the case for a profit-based utility function. Each element of the user space U can be defined with a profile that includes various user characteristics, such as age, gender, income, marital status, etc. In the simplest case, the profile can contain only a single (unique) element, such as User ID.

Similarly, each element of the item space I is defined via a set of characteristics. The central problem of RS lies in that a utility function f is usually not defined on the entire $U \times I$ space, but only on some subset of it. This means that f needs to be *generalized* to the entire space $U \times I$. In RS, a utility is typically represented by ratings and is initially defined only on the items previously rated by the users.

Generalizations from known to unknown ratings are usually done by:

- specifying heuristics that define the utility function and empirically validating its performance, or
- estimating the utility function that optimizes a certain performance criterion, such as Mean Absolute Error (MAE).

Once the unknown ratings are estimated, actual recommendations of an item to a user are made by selecting the highest rating among all the estimated ratings for that user, according to Eq. 12 . Alternatively, we can recommend the N best items to a user. Additionally, we can recommend a set of users to an item.

1.1.1 The Input to a Recommender System

The input to a RS depends on the type of the filtering algorithm employed. The input belongs to one of the following categories:

1. Ratings (also called votes), which express the opinion of users on items. Ratings are normally provided by the user and follow a specified numerical scale (example: 1-bad to 5-excellent). A common rating scheme is the binary rating scheme, which allows only ratings of either 0 or 1. Ratings can also be gathered implicitly from the user's purchase history, web logs, hyper-link visits, browsing habits or other types of information access patterns.
2. Demographic data, which refer to information such as the age, the gender and the education of the users. This kind of data is usually difficult to obtain. It is normally collected explicitly from the user.
3. Content data, which are based on content analysis of items rated by the user. The features extracted via this analysis are used as input to the filtering algorithm in order to infer a user profile.

1.1.2 The Output of a Recommender System

The output of a RS can be either a *prediction* or a *recommendation*.

- A *prediction* is expressed as a numerical value, $R_{a,j} = R(u_a, i_j)$, which represents the anticipated opinion of active user u_a for item i_j. This predicted value should necessarily be within the same numerical scale (example: 1-bad to 5-excellent) as the input referring to the opinions provided initially by active user u_a. This form of RS output is also known as *Individual Scoring*.
- A *recommendation* is expressed as a list of N items, where $N \leq n$, which the active user is expected to like the most. The usual approach in that case requires this list to include only items that the active user has not already purchased, viewed or rated. This form of RS output is also known as *Top- N Recommendation* or *Ranked Scoring*.

1.2 Methods of Collecting Knowledge About User Preferences

To generate personalized recommendations that are tailored to the specific needs of the active user, RS collect ratings of items by users and build user-profiles in ways that depend on the methods that the RS utilize to collect personal information about user preferences. In general, these methods are categorized into three approaches:

- an *Implicit approach*, which is based on recording user behavior,
- an *Explicit approach*, which is based on user interrogation,
- a *Mixing approach*, which is a combination of the previous two.

1.2.1 The Implicit Approach

This approach does not require active user involvement in the knowledge acquisition task, but, instead, the user behavior is recorded and, specifically, the way that he/she reacts to each incoming piece of data. The goal is to learn from the user reaction about the relevance of the data item to the user. Typical examples for implicit ratings are purchase data or reading time of Usenet news [4]. In the CF system in [8], they monitored reading times as an indicator for relevance. This revealed a relationship between time spent on reviewing data items and their relevance. In [9], the system learns the user profile by passively observing the hyperlinks clicked on and those passed over and by measuring user mouse and scrolling activity in addition to user browsing activity. Also, in [10] they utilize agents that operate as adaptive Web site RS. Through analysis of Web logs and web page structure, the agents infer knowledge of the popularity of various documents as well as a combination of document similarity. By tracking user actions and his/her acceptance of the agent recommendations, the agent can make further estimations about future recommendations to the specific user. The main benefits of implicit feedback over explicit ratings are that they remove the cognitive cost of providing relevance judgements explicitly and can be gathered in large quantities and aggregated to infer item relevance [11].

However, the implicit approach bears some serious implications. For instance, some purchases are gifts and, thus, do not reflect the active user interests. Moreover, the inference that purchasing implies liking does not always hold. Owing to the difficulty of acquiring explicit ratings, some providers of product recommendation services adopt bilateral approaches. For instance, www.Amazon.com computes recommendations based on explicit ratings whenever possible. In case of unavailability, observed implicit ratings are used instead.

1.2.2 The Explicit Approach

Users are required to explicitly specify their preference for any particular item, usually by indicating their extent of appreciation on 5-point or 7-point **Thurstone scales**. These scales are mapped to numeric values, e.g. $R_{i,j} \in [1, 2, 3, 4, 5]$. Lower values commonly indicate least favorable preferences, while higher values express the user's liking.[1] Explicit ratings impose additional efforts on users. Consequently, users often tend to avoid the burden of explicitly stating their preferences and either leave the system or rely upon "free-riding" [12]. Ratings made on these scales allow these judgments to be processed statistically to provide averages, ranges, or distributions. A central feature of explicit ratings is that the user who evaluates items has to examine them and, then, to assign to them values from the rating scale. This imposes a cognitive cost on the evaluator to assess the performance of an object [13].

1.2.3 The Mixing Approach

Newsweeder [14], a Usenet filtering system, is an example of a system that uses a combination of the explicit and the implicit approach, as it requires minimum user involvement. In this system, the users are required to rate documents for their relevance. The ratings are used as training examples for a machine learning algorithm that is executed nightly to generate user interest profiles for the next day.

---

[1] The Thurstone scale was used in psychology for measuring an attitude. It is made up of statements about a particular issue. A numerical value is associated with each statement, indicating how favorable or unfavorable the statement is judged to be.

Newsweeder is successful in reducing user involvement. However, the batch profiling used in Newsweeder is a shortcoming as profile adaptation is delayed significantly.

2 Summarization of Approaches to Recommendation

In general, the methods implemented in RS fall within one of the following categories:

- Content-based Methods
- Collaborative Methods
- Hybrid Methods.

2.1 Content-Based Methods

Modern information systems embed the ability to monitor and analyze users' actions to determine the best way to interact with them. Ideally, each user's actions are logged separately and analyzed to generate an individual user profile. All the information about a user, extracted either by monitoring user actions or by examining the objects the user has evaluated [15], is stored and utilized to customize services offered. This user modeling approach is known as *content-based learning*. The main assumption behind it is that a user's behavior remains unchanged through time; therefore, the content of past user actions may be used to predict the desired content of their future actions [16, 17]. Therefore, in content-based recommendation methods, the rating $R(u, i)$ of the item i for the user u is typically estimated based on ratings assigned by user u to the items $I_n \in I$ that are "similar" to item i in terms of their content, as defined by their associated features.

To be able to search through a collection of items and make observations about the similarity between objects that are not directly comparable, we must transform raw data at a certain level of information granularity. Information granules refer to a collection of data that contain only essential information. Such granulation allows more efficient processing for extracting features and computing numerical representations that characterize an item. As a result, the large amount of detailed information of one item is reduced to a limited set of features. Each feature is a vector of low dimensionality, which captures some aspects of the item and can be used to determine item similarity. Therefore, an item i could be described by a feature vector

$$F(i) = [feature_1(i), feature_2(i), feature_3(i), \ldots feature_n(i)]. \tag{3}$$

For example, in a music recommendation application, in order to recommend music files to user u, the content-based RS attempts to build a profile of the user's preferences based on features presented in music files that the user u has rated with high rating degrees. Consequently, only music files that have a high degree of similarity with these highly rated files would be recommended to the user. This method is known as "item-to-item correlation" [18]. The type of user profile derived by a content-based RS depends on the learning method which is utilized by the system. This approach to the recommendation process has its roots in information retrieval and information filtering [19, 20]. Retrieval-based approaches utilize interactive learning techniques such as *relevance feedback* methods, in order to organize and retrieve data in an effective personalized way. In relevance feedback methods, the user is part of the item-management process, which means that the user evaluates the results provided by the system. Then, the system adapts, its performance according to the user's preferences. In this way, the method of relevance feedback has the efficiency not only to take into account the user subjectivity in perceiving the content of items, but also to eliminate the gap between high-level semantics and low-level features which are usually used for the content description of items [21–23].

Besides the heuristics that are based mostly on information retrieval methods [19–23] such as the Rocchio algorithm or correlation-based schemes, other techniques for content-based recommendation utilize Pattern Recognition/Machine Learning approaches, such as Bayesian classifiers [24], clustering methods, decision trees, and artificial neural networks.

These techniques differ from information retrieval-based approaches as they calculate utility predictions based not on a heuristic formula, such as a cosine similarity measure, but rather are based on a model learnt from the underlying data using statistical and machine learning techniques. For example, based on a set of Web pages that were rated by the user as "relevant" or "irrelevant,", the naive Bayesian classifier is used in [24] to classify unrated Web pages.

Some examples of content-based methods come from the area of music data. In [25–29], they recommend pieces that are similar to users' favorites in terms of music content such as mood and rhythm. This allows a rich artist variety and various pieces, including unrated ones, to be recommended. To achieve this, it is necessary to associate user preferences with music content by using a practical database where most users tend to rate few pieces as favorites.

A relevance feedback approach for music recommendation was presented in [25] and based on the Tree Q vector quantization process initially proposed by Foote [30]. More specifically, relevance feedback was incorporated into the **user model** by modifying the quantization weights of desired vectors. Also, a relevance feedback music retrieval system, based on SVM Active Learning, was presented in [26], which retrieves the desired music piece according to mood and style similarity.

In [31], the authors explore the relation between the user's rating input, musical pieces with high degree of rating that were defined as the listener's favorite music, and music features. Specifically, labeled music pieces from specific artists were analyzed in order to build a correlation between user ratings and artists through music features. Their system forms the user profile as preference for music pieces of a specific artist. They confirmed that favorite music pieces were concentrated along certain music features.

The system in [32] proposes the development of a user-driven similarity function by combining timbre-, tempo-, genre-, mood-, and year-related features into the overall similarity function. More specifically, similarity is based on a weighted combination of these features and the end-user can specify his/her personal definition of similarity by weighting them.

The work in [33] tries to extend the use of signal approximation and characterization from genre classification to recognition of user taste. The idea is to learn music preferences by applying instance-based classifiers to user profiles. In other words, this system does *not* build an individual profile for every user, but instead tries to recognize his/her favorite genre by applying instance-based classifiers to user rating preferences by his/her music playlist.

2.2 Collaborative Methods

CF methods are based on the assumption that similar users prefer similar items or that a user expresses similar preferences for similar items. Instead of performing content indexing or content analysis, CF systems rely entirely on interest ratings from the members of a participating community [34]. CF methods are categorized into two general classes, namely *model-based* and *memory-based* [2, 7].

Model-based algorithms use the underlying data to learn a probabilistic model, such as a cluster model or a Bayesian network model [2, 35], using statistical and machine learning techniques. Subsequently, they use the model to make predictions. The clustering model [3, 36] works by clustering similar users in the same class and estimating the probability that a particular user is in a particular class. From there, the clustering model computes the conditional probability of ratings.

Memory-based methods, store raw preference information in computer memory and access it as needed to find similar users or items and to make predictions. In [10], CF was formulated as a classification problem. Specifically, based on a set of user ratings about items, they try to induce a model for each user that would allow the classification of unseen items into two or more classes, each of which corresponds to different points in the accepted rating scale.

Memory-based CF methods can be further divided into two groups, namely user-based and item-based [37] methods. On the one hand, user-based methods look for users (also called "neighbors") similar to the active user and calculate a predicted rating as a weighted average of the neighbor's ratings on the desired item. On the other hand, item-based methods look for similar items for an active user.

2.2.1 User-Based Collaborative Filtering Systems

User-based CF systems are systems that utilize **memory-based algorithms**, meaning that they operate over the entire user-item matrix R, to make predictions.

The majority of such systems mainly deal with **user-user similarity calculations**, meaning that they utilize user neighborhoods, constructed as collections of similar users. In other words, they deal with the rows of the user-item matrix, R, in order to generate their results. For example, in a personalized music RS called RINGO [38], similarities between the tastes of different users are utilized to recommend music items. This user-based CF approach works as follows: A new user is matched against the database to discover neighbors, which are other customers who, in the past, have had a similar taste as the new user, i.e. who have bought similar items as the new user. Items (unknown to the new user) that these neighbors like are then recommended to the new user. The main steps of this process are:

1. Representation of Input data,
2. Neighborhood Formation, and
3. Recommendation Generation.

1. Representation of Input Data To represent input data, one needs to define a set of ratings of users into a user-item matrix, R, where each $R(u, i)$ represents the rating value assigned by the user u to the item i. As users are not obligated to provide their opinion for all items, the resulting user-item matrix may be a **sparse matrix**. This sparsity of the user-item matrix is the main reason causing filtering algorithms not to produce satisfactory results. Therefore, a number of techniques were proposed to reduce the sparsity of the initial user-item matrix to improve the efficiency of the RS. *Default Voting* is the simplest technique used to reduce sparsity. A default rating value is inserted to items for which there does not exist a rating value. This rating value is selected to be neutral or somewhat indicative of negative preferences for unseen items [2].

An extension of the method of Default Voting is to use either the *User Average Scheme* or the *Item Average Scheme* or the *Composite Scheme* [39]. More specifically:

- In the *User Average Scheme*, for each user, u, the average user rating over all the items is computed, $\overline{R}(u)$. This is expressed as the average of the corresponding row in the user-item matrix. The user average is then used to replace any missing $R(u, i)$ value. This approach is based on the idea that a user's rating for a new item could be simply predicted if we take into account the same user's past ratings.
- In the *Item Average Scheme*, for each item, the item average over all users is computed, $\overline{R}(i)$. This is expressed as the average of the corresponding column in the user-item matrix. The item average is then used as a fill-in for missing values $R(u, i)$ in the matrix.
- In the *Composite Scheme*, the collected information for items and users both contribute to the final result. The main idea behind this method is to use the average of user u on item i as a base prediction and then add a correction term to it based on how the specific item was rated by other users.

  The scheme works as follows: When a missing entry regarding the rating of user u on item i is located, initially, the user average $\overline{R}(u)$ is calculated as the average of the corresponding user-item matrix row. Then, we search for existing ratings in the column which correspond to item i. Assuming that a set of l users, $U = \{u_1, u_2, \ldots, u_l\}$, has provided a rating for item i, we can compute a correction term for each user $u \in L$ equal to $\delta_k = R(u_k, i) - \overline{R}(u_k)$. After the corrections for all users in U are computed, the composite rating can be calculated as:

$$
R(u, i) = \begin{cases} \overline{R}(u) + \frac{\sum_{k=1}^{l} \delta_k}{l}, & \text{if } \text{user } u \text{ has not rated item } i \\ R, & \text{if } \text{user } u \text{ has rated item } i \text{ with } R. \end{cases} \tag{4}
$$

An alternative way of utilizing the composite scheme is through a simple transposition: first compute the item average, $\overline{R}(i_k)$, (i.e., average of the column which corresponds to item i) and then compute the correction terms, δ_k, by scanning through all l items $I = \{i_1, i_2, \ldots, i_l\}$ rated by user k. The fill-in value of $R(u, i)$ would then be:

$$
R(u, i) = \overline{R}(i) + \frac{\sum_{k=1}^{l} \delta_k}{l}, \tag{5}
$$

where l is the count of items rated by user u and the correction terms are computed for all items in I as $\delta_k = R(u, i_k) - \overline{R}(i_k)$

After generating a reduced-dimensionality matrix, we could use a vector similarity metric to compute the proximity between users and hence to form *neighborhoods of users* [40], as discussed in the following.

2. Neighborhood Formation In this step of the recommendation process, the *similarity* between users is calculated in the user-item matrix, R, i.e., users similar to the active user, u_a, form a proximity-based neighborhood with him. More specifically, neighborhood formation is implemented in two steps: Initially, the similarity between all the users in the user-item matrix, R, is calculated with the help of some proximity metrics. The second step is the actual neighborhood generation for the active user, where the similarities of users are processed in order to select those users that will constitute the neighborhood of the active user. To find the similarity between users u_a and u_b, we can utilize the *Pearson correlation metric*. The Pearson correlation was initially introduced in the context of the GroupLens project [4, 38], as follows: Let us assume that a set of m users u_k, where $k = 1, 2, \ldots, m$, $U_m = \{u_1, u_2, \ldots, u_m\}$, have provided a rating $R(u_k, i_l)$ for item i_l, where $l = 1, 2, \ldots, n$, $I_n = \{i_1, i_2, \ldots, i_n\}$ is the set of items. The Pearson correlation coefficient is given by:

$$
sim(u_a, u_b) = \frac{\sum_{l=1}^{n} (R(u_a, i_l) - \overline{R}(u_a))(R(u_b, i_l) - \overline{R}(u_b))}{\sqrt{\sum_{l=1}^{n} (R(u_a, i_l) - \overline{R}(u_a))^2 \sum_{l=1}^{n} (R(u_b, i_l) - \overline{R}(u_b))^2}}. \tag{6}
$$

Another metric similarity uses the *cosine-based approach* [2], according to which the two users u_a and u_b, are considered as two vectors in *n-dimensional* item-space, where $n = |I_n|$. The similarity between two vectors can be measured by computing the cosine angle between them:

$$sim(u_a, u_b) = \cos(\overrightarrow{u_a}, \overrightarrow{u_b}) = \frac{\sum_{l=1}^{n} R(u_a, i_l) R(u_b, i_l)}{\sqrt{\sum_{l=1}^{n} R(u_a, i_l)^2} \sqrt{\sum_{l=1}^{n} R(u_b, i_l)^2}}. \qquad (7)$$

In RS, the use of the Pearson correlation similarity metric to estimate the proximity among users performs better than the cosine similarity [2].

At this point in the recommendation process, a single user is selected who is called the *active user*. The active user is the user for whom the RS will produce predictions and proceed with generating his/her neighborhood of users. A *similarity matrix S* is generated, containing the similarity values between all users. For example, the ith row in the similarity matrix represents the similarity between user u_i and all the other users. Therefore, from this similarity matrix S various schemes can be used in order to select the users that are most similar to the active user. One such scheme is the *center-based scheme*, in which from the row of the active user u_a are selected those users who have the highest similarity value with the active user.

Another scheme for neighborhood formation is the *aggregate neighborhood formation scheme*. In this scheme, a neighborhood of users is created by finding users who are closest to the *centroid* of the current neighborhood and not by finding the users who are closest to the active user himself/herself. This scheme allows all users to take part in the formation of the neighborhood, as they are gradually selected and added to it.

3. Generation of Recommendations The generation of recommendations is represented by predicting a rating, i.e., by computing a numerical value which constitutes a predicted opinion of the active user u_a for an item i_j unseen by him/her. This predicted value should be within the same accepted numerical scale as the other ratings in the initial user-item matrix R. In the generation of predictions, only those users participate that lie within the neighborhood of the active user. In other words, only a subset of k users participate from the m users in the set U_m that have provided ratings for the specific item i_j, $U_k \subseteq U_m$. Therefore, a *prediction score* P_{u_a, i_j} is computed as follows [4]:

$$P_{u_a, i_j} = \overline{R}(u_a) + \frac{\sum_{t=1}^{k} (R(u_t, i_j) - \overline{R}(u_t)) * sim(u_a, u_t)}{\sum_{t=1}^{k} |sim(u_a, u_t)|}, \qquad \text{where } U_k \subseteq U_l \quad (8)$$

Here, $\overline{R}(u_a)$ and $\overline{R}(u_t)$ are the average rating of the active user u_a and u_t, respectively, while $R(u_t, i_j)$ is the rating given by user u_t to item i_j. Similarity $sim(u_a, u_t)$ is the similarity among users u_a and u_t, computed using the Pearson correlation in Eq. 6. Finally, the RS will output several items with the best predicted ratings as the recommendation list.

An alternative output of a RS is the *top-N recommendations* output. In this case, recommendations form a list of N items that the active user is expected to like the most. For the generation of this list, users are ranked first according to their similarity to the active user. The k most similar (i.e. most highly ranked) users are selected as the k-nearest neighbors of the active user u_a. The frequency count of an item is calculated by scanning the rating of the item by the k-nearest neighbors. Then, the items are sorted based on frequency count. The N most frequent items that have not been rated by the active user are selected as the top-N recommendations [41].

2.2.2 Item-Based Collaborative Filtering Systems

A different approach [5, 37] is based on item relations and not on user relations, as in classic CF. Since the relationships between users are relatively dynamic, as they continuously buy new products, it is computationally hard to calculate the user-to-user matrix online. This causes the user-based CF approach to be relatively expensive in terms of computational load. In the item-based CF algorithm, we look into the set of items, denoted by I_{u_a}, that the active user, u_a, has rated and compute how similar they are to the target item i_t. Then, we select the k most similar items $I_k = \{i_1, i_2, \ldots, i_k\}$, based on their corresponding similarities $\{sim(i_t, i_1), sim(i_t, i_2), \ldots, sim(i_t, i_k)\}$. The predictions can then be computed by taking a weighted average of the active user's ratings on these similar items. The main steps in this approach are the same as in user-based CF. The difference in the present approach is that instead of calculating similarities between two users who have provided ratings for a common item, we calculate similarities between two items i_t, i_j which have been rated by a common user u_a. Therefore, the Pearson correlation coefficient and cosine similarity are, respectively, given as:

$$sim(i_t, i_j) = \frac{\sum_{l=1}^{n} (R(u_l, i_t) - \overline{R}(i_t))(R(u_l, i_j) - \overline{R}(i_j))}{\sqrt{\sum_{l=1}^{n} (R(u_l, i_t) - \overline{R}(i_t))^2 \sum_{l=1}^{n} (R(u_l, i_j) - \overline{R}(i_j))^2}} \qquad (9)$$

$$sim(i_t, i_j) = \cos(\overrightarrow{i_t}, \overrightarrow{i_j}) = \frac{\sum_{l=1}^{n} R(u_l, i_t) R(u_l, i_j)}{\sqrt{\sum_{l=1}^{n} R(u_l, i_t)^2} \sqrt{\sum_{l=1}^{n} R(u_l, i_j)^2}}. \qquad (10)$$

Next, the similarities between all items in the initial user-item matrix, R, are calculated. The final step in the CF procedure is to isolate k items from n, $(I_k \subseteq I_n)$ in order to share the greatest similarity with item i_t for which we are seeking a prediction, form its neighborhood of items, and proceed with prediction generation. A prediction on item i_t for active user u_a is computed as the sum of ratings given by the active user on items belonging to the neighborhood I_k. These ratings are weighted by the corresponding similarity, $sim(i_t, i_j)$ between item i_t and item i_j, with $j = 1, 2, \ldots, k$, taken from neighborhood I_k:

$$P_{u_a, i_j} = \frac{\sum_{j=1}^{k} sim(i_t, i_j) * R(u_a, i_j)}{\sum_{j=1}^{k} |sim(i_t, i_j)|} \qquad \text{where } I_k \subseteq I_n. \qquad (11)$$

In [42], the authors proposed that the long-term interest profile of a user (*task profile*) be established either by explicitly providing some items associated with the current task or by implicitly observing the user behavior (*intent*). By utilizing the item-to-item correlation matrix, items that resemble the items in the task profile are selected for recommendation. Since they match the task profile, these items fit the current task of the user. Before recommending them to the user, these items will be re-ranked to fit the user interests based on the interest prediction.

2.2.3 Personality Diagnosis

Personality diagnosis may be thought of as a hybrid between memory and model-based approaches of CF. The main characteristic is that predictions have meaningful probabilistic semantics. Moreover, this approach assumes that preferences constitute a characterization of their underlying personality type for each user. Therefore, taking into consideration the active user's known ratings of items, it is possible to estimate the probability that he/she has the same personality type with another user. The personality type of a given user is taken to be the vector of "true" ratings for items the user has seen. A true rating differs from the actually reported rating given by a user by an amount of (Gaussian) noise. Given the personality type of a user, the personality diagnosis approach estimates the probability that the given user is of the same personality type as other users in the system, and, consequently, estimates the probability that the user will like some new item [43].

The personality type for each user u_k is formulated as follows, where $k = 1, 2, \ldots, m$, $U_m = \{u_1, u_2, \ldots, u_m\}$, and the user u_k has a number of preferred items in $I_n = \{i_1, i_2, \ldots, i_n\}$:

$$\overset{true}{R}(u_k) = \left\{ \overset{true}{R}(u_k, i_1), \overset{true}{R}(u_k, i_2), \ldots, \overset{true}{R}(u_k, i_n) \right\}. \qquad (12)$$

Here, $\overset{true}{R}(u_k, i_l)$, with $i_l \in I_n$ and $l = 1, 2, \ldots, n$, stands for true rating by user u_k of the item i_l. It is important to note the difference between *true* and *reported* (given) ratings of the user. The true ratings encode the underlying internal preferences for a user that are *not* directly accessible by the designer of the RS. However, the reported ratings are those which were provided by users and utilized by the RS.

It is assumed that the reported ratings given by users include Gaussian noise. This assumption has the meaning that one user could report different ratings for the same items under different situations, depending on the context. Thus, we can assume that the rating reported by the user for an item i_l is drawn from an independent normal distribution with mean $\overset{true}{R}(u_k, i_l)$. Particularly:

$$Pr\left(R(u_k, i_l) = x \middle| \overset{true}{R}(u_k, i_l) = y\right) \propto e^{-\frac{(x-y)^2}{2\sigma^2}}, \tag{13}$$

where σ is a free parameter, x is the rating that the user has reported to the RS, and y is the true rating value that the user u_k would have reported if there no noise were present.

Furthermore, we assume that the distribution of personality types in the rating array R of users-items is representative of the personalities found in the target population of users. Therefore, taking into account this assumption, we can formulate the prior probability $Pr\left(\overset{true}{R}(u_a) = v\right)$ that the active user u_a rates items accordingly to a vector v as given by the frequency that the other users rate according to v. Thereby, instead of explicitly counting occurrences, we simply define $\overset{true}{R}(u_a)$ to be a random variable that can take one of m values, $(R(u_1), R(u_2), \ldots, R(u_m))$, each with probability $\frac{1}{m}$:

$$Pr\left(\overset{true}{R}(u_a) = R(u_k)\right) = \frac{1}{m}. \tag{14}$$

Combining Eqs. 13 and 14 and given the active user's ratings, we can compute the probability that the active user is of the same personality type as any other user, by applying the Bayes rule:

$$Pr\left(\overset{true}{R}(u_a) = R(u_k) \middle| R(u_a, i_1) = x_1, \ldots, R(u_a, i_n) = x_n\right)$$
$$\propto Pr\left(R(u_a, i_1) = x_1 \middle| \overset{true}{R}(u_a, i_1) = R(u_a, i_1)\right)$$
$$\ldots Pr\left(R(u_a, i_n) = x_n \middle| \overset{true}{R}(u_a, i_n) = R(u_a, i_n)\right)$$
$$\cdot Pr\left(\overset{true}{R}(u_a) = R(u_k)\right). \tag{15}$$

Hence, computing this quantity for each user u_k, we can compute the probability distribution for the active user's rating of an unseen item i_j. This probability distribution corresponds to the prediction P_{u_a, i_j} produced by the RS and equals the expected rating value of active user u_a for the item i_j:

$$P_{u_a,i_j} = Pr\big(R(u_a,i_j) = x_j | R(u_a,i_1) = x_1, \ldots, R(u_a,i_n) = x_n\big)$$

$$= \sum_{k=1}^{m} Pr\left(R(u_a,i_j) = x_j \Big| \overset{true}{R}(u_a) = R(u_k)\right) \qquad (16)$$

$$\cdot Pr\left(\overset{true}{R}(u_a) = R(u_k) | R(u_a,i_1) = x_1, \ldots, R(u_a,i_n) = x_n\right).$$

The model is depicted as a naive Bayesian network with the structure of a classical diagnostic model as follows:

- Firstly, we observe ratings and, using Eq. 15, compute the probability that each personality type is the cause. Ratings can be considered as "symptoms" while personality types as "diseases" leading to those symptoms in the diagnostic model.
- Secondly, we can compute the probability of rating values for an unseen item using Eq. 16. The most probable rating is returned as the prediction of the RS.

An alternative interpretation of personality diagnosis is to consider it as a clustering method with exactly one user per cluster. This is so because each user corresponds to a single personality type and the effort is to assign the active user to one of these clusters [2, 3].

An additional interpretation of personality diagnosis is that the active user is assumed to be "generated" by choosing one of the other users uniformly at random and adding Gaussian noise to his/her ratings. Given the active user's known ratings, we can infer the probability that he/she be actually one of other users and then compute probabilities for ratings of other items.

2.3 Hybrid Methods

Hybrid methods combine two or more recommendation techniques to achieve better performance and to take out drawbacks of each technique separately. Usually, CF methods are combined with content-based methods. According to [7], hybrid RS could be classified into the following categories:

- Combining Separate Recommenders
- Adding Content-Based Characteristics to Collaborative Models
- Adding Collaborative Characteristics to Content-Based Models
- A Single Unifying Recommendation Model.

Combining Separate Recommenders The Hybrid RS of this category include two separate systems, a collaborative one and a content-based one. There are four different ways of combining these two separate systems, namely the following:

- *Weighted Hybridization Method.* The outputs (ratings) acquired by individual RS are combined together to produce a single final recommendation using either a linear combination [44] or a voting scheme [10]. The *P-Tango* system [44] initially gives equal weights to both recommenders, but gradually adjusts the weights as predictions about user ratings are confirmed or not. The system keeps the two filtering approaches separate and this allows the benefit from individual advantages.
- *Switched Hybridization Method.* The system switches between recommendation techniques selecting the method that gives better recommendations for the current situation depending on some recommendation "quality" metric. A characteristic example of such a recommender is *The Daily Learner* [45], which selects the recommender sub-system that provides the higher level of confidence. Another example of this method is presented in [46] where either the content-based or the collaborative filtering technique is selected according to which of the two provided better consistency with past ratings of the user.
- *Mixed Hybridization Method.* In this method, the results from different recommender sub-systems are presented simultaneously. An example of such a recommender is given in [47] where they utilize a content-based technique based on textual descriptions of TV shows and collaborative information about users' preferences. Recommendations from both techniques are provided together in the final suggested program.
- *Cascade Hybridization Method.* In this method, one recommendation technique is utilized to produce a coarse ranking of candidates, while the second technique focuses only on those items for which additional refinement is needed. This method is more efficient than the weighted hybridization method which applies all of its techniques on all items. The computational burden of this hybrid approach is relatively small because recommendation candidates in the second level are partially eliminated in the first level. Moreover this method is more tolerant to noise in the operation of low-priority recommendations, since ratings of the high level recommender can only be refined, but never over-turned [15]. In other words, cascade hybridization methods can be analyzed into two sequential stages. The first stage (content-based method or knowledge-based/ collaborative) selects intermediate recommendations. Then, the second stage (collaborative/content-based method or knowledge-based) selects appropriate items from the recommendations of the first stage. Burke [48] developed a restaurant RS called *EntreeC*. The system first selects several restaurants that match a user's preferred cuisine (e.g., Italian, Chinese, etc.) with a knowledge-based method. In the knowledge-based method, the authors construct a feature vector according to defined attributes that characterize the restaurants. This method is similar to content-based methods; however, it must be noted that these metadata are content-independent and for this reason the term *knowledge-based* is utilized. These restaurants are then ranked with a collaborative method.

2.3.1 Adding Content-Based Characteristics to Collaborative Models

In [10], the authors proposed *collaboration via content*. This is a method that uses a prediction scheme similar to the standard CF, in which similarity among users is not computed on provided ratings, but rather on the content-based profile of each user. The underlying intuition is that like-minded users are likely to have similar content-based models and that this similarity relation can be detected without requiring overlapping ratings. The main limitation of this approach is that the similarity of users is computed using Pearson's correlation coefficient between content-based weight vectors.

On the other hand, in [49] the authors proposed the *content-boosted collaborative filtering* approach, which exploits a content-based predictor to enhance existing user data and then provides personalized suggestions through CF. The content-based predictor is applied to each row of the initial user-item matrix, corresponding to each user, and gradually generates a pseudo user-item matrix that is a full dense matrix. The similarity between the active user, u_a, and another user, u_i, is computed with CF using the new pseudo user-item matrix.

2.3.2 Adding Collaborative Characteristics to Content-Based Models

The main technique of this category is to apply dimensionality reduction on a group of content-based profiles. In [50], the authors used *latent semantic indexing* to create a collaborative view of a collection of user profiles represented as term vectors. This technique results in performance improvement in comparison with the pure content-based approach.

2.3.3 A Single Unifying Recommendation Model

A general unifying model that incorporates content-based and collaborative characteristics was proposed in [36], where the authors present the use of content-based and collaborative characteristics (e.g., the age or gender of users or the genre of movies) in a single rule-based classifier. Single unifying models were also presented in [51], where the authors utilized a unified probabilistic method for combining collaborative and content-based recommendations.

2.3.4 Other Types of Recommender Systems

Demographics-based RS. The basis for recommendations in demographics-based RS is the use of prior knowledge on demographic information about the users and their opinions for the recommended items. Demographics-based RS classify their users according to personal demographic data (e.g. age and gender) and classify items into user classes. Approaches falling into this group can be found in Grundy

[52], a system for book recommendation, and in [53] for marketing recommendations. Similarly to CF, demographic techniques also employ user-to-user correlations, but differ in the fact that they do not require a history of user ratings. An additional example of a demographics-based RS is described in [10], in which information about users is taken from their home-pages to avoid the need to maintain a history of user ratings. Demographic characteristics for users (e.g. their age and gender) is also utilized in [36].

Knowledge-based RS. Knowledge-based RS use prior knowledge on how the recommended items fulfill the user needs. Thus, the goal of a knowledge-based RS is to reason about the relationship between a need and a possible recommendation. The user profile should encompass some knowledge structure that supports this inference. An example of such a RS is presented in [48], where the system *Entree* uses some domain knowledge about restaurants, cuisines, and foods to recommend a restaurant to its users. The main advantage using a knowledge-based system is that there is no bootstrapping problem. Because the recommendations are based on prior knowledge, there is no learning time before making good recommendations. However, the main drawback of knowledge-based systems is a need for knowledge acquisition for the specific domain which makes difficult the adaptation in another domain and not easily adapted to the individual user as it is enhanced by predefined recommendations.

2.4 Fundamental Problems of Recommender Systems

Cold Start Problem. The cold-start problem [54] is related to the learning rate curve of a RS. The problem could be analyzed into two different sub-problems:

- **New-User Problem**, i.e., the problem of making recommendations to a new user [55], where almost nothing is known about his/her preferences.
- **New-Item Problem**, i.e., the problem where ratings are required for items that have not been rated by users. Therefore, until the new item is rated by a satisfactory number of users, the RS would not be able to recommend this item. This problem appears mostly in collaborative approaches and could be eliminated with the use of content-based or hybrid approaches where content information is used to infer similarities among items.

  This problem is also related, with the *coverage* of a RS, which is a measure for the domain of items over which the system could produce recommendations. For example, low coverage of the domain means that only a limited space of items is used in the results of the RS and these results usually could be biased by preferences of other users. This is also known as the problem of *over-specialization*. When the system can only recommend items that score highly against a user's profile, the user is limited to being recommended items that are similar to those already rated. This problem, which has also been studied in other domains, is often addressed by introducing some randomness. For example, the use of genetic algorithms has been proposed as a possible solution in the context of information filtering [56].

Novelty Detection—Quality of Recommendations. From those items that a RS recommends to users, there are items that are already known to the users and items that are new (novel) and unknown to them. Therefore, there is a competitiveness between the desire for novelty and the desire for high quality recommendations. On one hand, the quality of the recommendations [40] is related to "trust" that users express for the recommendations. This means that a RS should minimize false positive errors and, more specifically, the RS should not recommend items that are not desirable. On the other hand, novelty is related with the "timestamp-age" of items: the older items should be treated as less relevant than the newer ones and this causes increase to the novelty rate. Thus, a high novelty rate will produce poor quality recommendations because the users will not be able to identify most of the items in the list of recommendations.

Sparsity of Ratings. The sparsity problem [7, 57] is related to the unavailability of a large number of rated items for each active user. The number of items that are rated by users is usually a very small subset of those items that are totally available. For example, in *Amazon*, if the active users may have purchased 1 % of the items and the total amount of items is approximately 2 millions of books, this means that there are only 20,000 of books which are rated. Consequently, such sparsity in ratings degrades the accurate selection of the neighbors in the step of neighborhood formation and leads to poor recommendation results.

A number of possible solutions have been proposed to overcome the sparsity problem such as content-based similarities, item-based CF methods, use of demographic data and a number of hybrid approaches [15]. A different approach to deal with this problem is proposed in [58], where the authors utilized *dimension reduction techniques*, such as singular value decomposition, in order to transform the sparse user-item matrix R into a dense matrix. The SVD is a method for matrix factorization that produces the best lower-rank approximations to the original matrix [10].

Scalability. RS, especially with large electronic sites, have to deal with a constantly growing number of users and items [2, 3]. Therefore, an increasing amount of computational resources is required as the amount of data grows. A recommendation method, that could be efficient when the number of data is limited, could be very time-consuming and scale poorly. Such a method would be unable to generate a satisfactory number of recommendations from a large amount of data. Thus, it is important that the recommendation approach be capable of scaling up in a successful manner [37].

Lack of Transparency Problem. RS are usually black boxes, which means that RS are not able to explain to their users why they recommend those specific items. In content-based approaches [29, 59], this problem could be minimized. However, in collaborative approaches, predictions may be harder to explain than predictions made by content-based models [60].

Gray Sheep User Problem. The majority of users falls into the class of so called "white-sheep", i.e. those who have high correlation with many other users. For these users, it should be easy to find recommendations. In a small or even medium community of users, there are users whose opinions do not consistently agree or disagree with any group of people [44]. There are users whose preferences are atypical (uncommon) and vary significantly from the norm. After neighborhood formation, these users will not have many other users as neighbors. As a result, there will be poor recommendations for them. From a statistical point of view, as the number of users of a system increases, so does the probability of finding other people with similar preferences, which means that better recommendations could be provided [61].

References

1. Resnick, P., Varian, H.R.: Recommender systems. Commun. ACM **40**(3), 56–57 (1997)
2. Breese, J.S., Heckerman, D., Kadie, C.: Empirical analysis of predictive algorithms for collaborative filtering. In: Proceedings of Fourteenth Conference on Uncertainty in Artificial Intelligence, pp. 43–52. Morgan Kaufmann, San Francisco (1998)
3. Ungar, L., Foster, D., Andre, E., Wars, S., Wars, F.S., Wars, D.S., Whispers, J.H.: Clustering methods for collaborative filtering. In: Proceedings of AAAI Workshop on Recommendation Systems. AAAI Press, Madison (1998)
4. Resnick, P., Iacovou, N., Suchak, M., Bergstrom, P., Riedl, J.: Grouplens: an open architecture for collaborative filtering of netnews. In: Proceedings of Computer Supported Collaborative Work Conference, pp. 175–186. ACM Press, Chapel Hill (1994)
5. Karypis, G.: Evaluation of item-based top-n recommendation algorithms. In: CIKM '01: Proceedings of the Tenth International Conference on Information and Knowledge Management, pp. 247–254. ACM, New York (2001)
6. Malone, T.W., Grant, K.R., Turbak, F.A., Brobst, S.A., Cohen, M.D.: Intelligent information-sharing systems. Commun. ACM **30**(5), 390–402 (1987). doi:10.1145/22899.22903
7. Adomavicius, G., Tuzhilin, E.: Toward the next generation of recommender systems: A survey of the state-of-the-art and possible extensions. IEEE Trans. Knowl. Data Eng. **17**, 734–749 (2005)
8. Konstan, J.A., Miller, B.N., Maltz, D., Herlocker, J.L., Gordon, L.R., Riedl, J.: GroupLens: applying collaborative filtering to Usenet news. Commun. ACM **40**(3), 77–87 (1997)
9. Jude, J.G., Shavlik, J.: Learning users' interests by unobtrusively observing their normal behavior. In: Proceedings of International Conference on Intelligent User Interfaces, pp. 129–132. ACM Press, New York (2000)
10. Pazzani, M.J.: A framework for collaborative, content-based and demographic filtering. Artif. Intell. Rev. **13**(5–6), 393–408 (1999)
11. Kelly, D., Teevan, J.: Implicit feedback for inferring user preference: a bibliography. SIGIR Forum **37**(2), 18–28 (2003)
12. Avery, C., Zeckhauser, R.: Recommender systems for evaluating computer messages. Commun. ACM **40**(3), 88–89 (1997)
13. Nichols, D.M.: Implicit rating and filtering. In: Proceedings of the Fifth DELOS Workshop on Filtering and Collaborative Filtering, pp. 31–36 (1997)
14. Lang, K.: Newsweeder: learning to filter netnews. In: Proceedings of 12th International Machine Learning Conference (ML95), pp. 331–339 (1995)

15. Burke, R.: Hybrid recommender systems: survey and experiments. User Model. User-Adapt. Interact. **12**(4), 331–370 (2002)
16. Mooney, R.J., Roy, L.: Content-based book recommending using learning for text categorization. In: DL '00: Proceedings of the Fifth ACM Conference on Digital Libraries, pp. 195–204. ACM, New York (2000)
17. Balabanović, M., Shoham, Y.: Fab: content-based, collaborative recommendation. Commun. ACM **40**(3), 66–72 (1997)
18. Schafer, J.B., Konstan, J.A., Riedl, J.: E-commerce recommendation applications. Data Min. Knowl. Discov. **5**(1–2), 115–153 (2001)
19. Baeza-Yates, R., Ribeiro-Neto, B. (eds.): Modern Information Retrieval. Addison-Wesley, Reading (1999)
20. Salton, G. (ed.): Automatic Text Processing. Addison-Wesley, Reading (1989)
21. Rui, Y., Huang, T.S., Ortega, M., Mehrotra, S.: Adaptive algorithms for interactive multimedia. IEEE Trans. Circ. Syst. Video Technol. **8**(5), 644–655 (1998)
22. Cox, I.J., Miller, M.L., Omohundro, S.M., Yianilos, P.N.: Pichunter: bayesian relevance feedback for image retrieval. Int. Conf. Pattern Recogn. **3**, 361 (1996)
23. Doulamis, N.D., Doulamis, A.D., Varvarigou, T.A.: Adaptive algorithms for interactive multimedia. IEEE MultiMedia **10**(4), 38–47 (2003)
24. Pazzani, M., Billsus, D.: Learning and revising user profiles: The identification of interesting web sites. Machine Learning **27**(3), 313–331 (1997)
25. Hoashi, K., Matsumo, K., Inoue, N.: Personalization of user profiles for content-based music retrieval based on relevance feedback. Presented at the Proceedings of ACM International Conference on Multimedia 2003, pp. 110–119. ACM press, New York (2003)
26. Mandel, M., Poliner, G., Ellis, D.: Support vector machine active learning for music retrieval ACM multimedia systems journal. ACM Multim. Syst. J. **12**(1), 3–13 (2006)
27. Logan, B.: Music recommendation from song sets. Presented at the Proceedings of 5th International Conference on Music Information Retrieval, pp. 425–428 (2004)
28. Celma, O., Ramrez, M., Herrera, P.: Foafing the music: a music recommendation system based on RSS feeds and user preferences. In: Proceedings of the 6th International Conference on Music Information Retrieval (ISMIR), London, UK (2005)
29. Sotiropoulos, D.N., Lampropoulos, A.S., Tsihrintzis, G.A.: Musiper: a system for modeling music similarity perception based on objective feature subset selection. User Model. User-Adapt. Interact. **18**(4), 315–348 (2008)
30. Foote, J.: An overview of audio information retrieval. Multimed. Syst. **7**(1), 2–10 (1999)
31. Arakawa, K., Odagawa, S., Matsushita, F., Kodama, Y., Shioda, T.: Analysis of listeners' favorite music by music features. Presented at the Proceedings of the International Conference on Consumer Electronics (ICCE), IEEE, pp. 427–428 (2006)
32. Vignoli, F., Pauws, S.: A music retrieval system based on user driven similarity and its evaluation. Presented at the Proceedings of 6th International Conference on Music Information Retrieval, London, UK, pp. 272–279, Sept 2005
33. Grimaldi, M., Cunningham, P.: Experimenting with music taste prediction by user profiling. Presented at the Proceedings of Music Information Retrieval (MIR'04), New York, USA, Oct 2004
34. Herlocker, J.L., Konstan, J.A., Riedl, J.: An empirical analysis of design choices in neighborhood-based collaborative filtering algorithms. Inf. Retr. **5**(4), 287–310 (2002)
35. Zhang, Y., Koren, J.: Efficient bayesian hierarchical user modeling for recommendation system. In: Proceedings of 30th Annual International ACM SIGIR Conference on Research and Development in Information Retrieval, pp. 47–54. ACM, New York, 2007
36. Basu, C., Hirsh, H., Cohen, W.: Recommendation as classification: using social and content-based information in recommendation. In: AAAI '98/IAAI '98: Proceedings of the Fifteenth National/Tenth Conference on Artificial Intelligence/Innovative Applications of Artificial Intelligence, pp. 714–720. American Association for Artificial Intelligence, Menlo Park, 1998

37. Sarwar, B., Karypis, G., Konstan, J., Reidl, J.: Item-based collaborative filtering recommendation algorithms. In: Proceedings of 10th International Conference on World Wide Web, pp. 285–295. ACM, New York, 2001
38. Shardanand, U., Maes, P.: Social information filtering: algorithms for automating "word of mouth". In: Proceedings of SIGCHI Conference on Human Factors in Computing Systems, pp. 210–217. ACM Press/Addison-Wesley, New York, 1995
39. Sarwar, B.M.: Scalability, and distribution in recommender systems. PhD Dissertation, University of Minnesota (2001)
40. Sarwar, B., Karypis, G., Konstan, J., Riedl, J.: Analysis of recommendation algorithms for e-commerce. In: Proceedings of 2nd ACM Conference on Electronic Commerce, pp. 158–167. ACM, New York, 2000
41. Liu, D.-R., Shih, Y.-Y.: Integrating ahp and data mining for product recommendation based on customer lifetime value. Inf. Manage. 42(3), 387–400 (2005)
42. Herlocker, J.L., Konstan, J.A.: Content-independent task-focused recommendation. IEEE Internet Comput. 5(6), 40–47 (2001)
43. Pennock, D.M., Horvitz, E., Lawrence, S., Giles, C.L.: Collaborative filtering by personality diagnosis: a hybrid memory and model-based approach. In: UAI '00: Proceedings of the 16th Conference on Uncertainty in Artificial Intelligence, pp. 473–480. Morgan Kaufmann, San Francisco, 2000
44. Claypool M., Gokhale, A., Miranda, T., Murnikov, P., Netes, D., Sartin, M.: Combining content-based and collaborative filters in an online newspaper. In: Proceedings of ACM SIGIR Workshop on Recommender Systems (1999)
45. Billsus, D., Pazzani, M.J.: User modeling for adaptive news access. User Model. User-Adapt. Interact. 10(2–3), 147–180 (2000)
46. Tran, T., Cohen, R.: Hybrid recommender systems for electronic commerce. Presented at the Proceedings of Knowledge-Based Electronic Markets, Papers from the AAAI Workshop, AAAI Technical Report WS-00-04, pp. 78–83, Menlo Park, CA (2000)
47. Smyth, B., Cotter, P.: A personalized TV listings service for the digital TV age. Knowl.-Based Syst. 13, 53–59 (2000)
48. Burke, R.: Knowledge-based recommender systems. In: Kent, A. (ed.) Encyclopedia of Library and Information Systems. Springer, New York (2000)
49. Melville, P., Mooney, R.J., Nagarajan, R.: Content-boosted collaborative filtering for improved recommendations. In: Proceedings of Eighteenth National Conference on Artificial Intelligence, pp. 187–192. American Association for Artificial Intelligence, Menlo Park, 2002
50. Soboroff, I., Nicholas, C., Nicholas, C.K.: Combining content and collaboration in text filtering. In: Proceedings of the IJCAI99 Workshop on Machine Learning for Information Filtering, pp. 86–91 (1999)
51. Popescul, A., Ungar, L.H., Pennock, D.M., Lawrence, S.: Probabilistic models for unified collaborative and content-based recommendation in sparse-data environments. In: UAI '01: Proceedings of the 17th Conference in Uncertainty in Artificial Intelligence, pp. 437–444. Morgan Kaufmann, San Francisco (2001)
52. Rich, E.: User Modeling via Stereotypes, pp. 329–342. Morgan Kaufmann, San Francisco (1998)
53. Krulwich, B.: Lifestyle finder: intelligent user profiling using large-scale demographic data. AI Magazine 18(2), 37–45 (1997)
54. Schein, A.I., Popescul, A., Ungar, L.H., Pennock, D.M.: Methods and metrics for cold-start recommendations. In: SIGIR '02: Proceedings of the 25th Annual International ACM SIGIR Conference on Research and Development in Information Retrieval, pp. 253–260. ACM, New York (2002)
55. Rashid, A.M., Albert, I., Cosley, D., Lam, S.K., McNee, S.M., Konstan, J.A., Riedl, J.: Getting to know you: learning new user preferences in recommender systems. In: IUI '02: Proceedings of the 7th International Conference on Intelligent User Interfaces, pp. 127–134. ACM, New York (2002)

56. Sheth, B., Maes, P.: Evolving agents for personalized information filtering. In: Proceedings of 19th conference on Artificial Intelligence for Applications, pp. 345–352 (1993)
57. Linden, G., Smith, B., York, J.: Amazon.com recommendations: item-to-item collaborative filtering. IEEE Internet Comput. **7**(1), 76–80 (2003)
58. Sarwar, B.M., Karypis, G., Konstan, J.A., Riedl, J.T.: Application of dimensionality reduction in recommender system—a case study. In: ACM WebKDD Workshop (2000)
59. Symeonidis, P., Nanopoulos, A., Manolopoulos, Y.: Providing justifications in recommender systems. IEEE Trans. Syst. Man Cybernet. Part A: Syst. Humans **38**(6), 1262–1272 (2008)
60. Herlocker, J.L., Konstan, J.A., Riedl, J.: Explaining collaborative filtering recommendations. In: CSCW '00: Proceedings of the 2000 ACM Conference on Computer Supported Cooperative Work, pp. 241–250. ACM, New York (2000)
61. Terveen, L., Hill, W.: Human-computer collaboration in recommender systems. In: Carroll, J. (ed.) HCI in the New Millennium. Addison Wesley, Reading (2001)

Hybrid User Model for Capturing a User's Information Seeking Intent

Hien Nguyen and Eugene Santos Jr.

Abstract A user is an important factor that contributes to the success or failure of any information retrieval system. Unfortunately, users often do not have the same technical and/or domain knowledge as the designers of such a system, while the designers are often limited in their understanding of a target user's needs. In this chapter, we study the problem of employing a cognitive user model for information retrieval in which knowledge about a user is captured and used for improving his/her performance in an information seeking task. Our solution is to improve the effectiveness of a user in a search by developing a *hybrid user model* to capture user intent dynamically and combines the captured intent with an awareness of the components of an information retrieval system. The term "hybrid" refers to the methodology of combining the understanding of a user with the insights into a system all unified within a decision theoretic framework. In this model, multi-attribute utility theory is used to evaluate values of the attributes describing a user's intent in combination with the attributes describing an information retrieval system. We use the existing research on predicting query performance and on determining dissemination thresholds to create functions to evaluate these selected attributes. This approach also offers fine-grained representation of the model and the ability to learn a user's knowledge dynamically. We compare this approach with the best traditional approach for relevance feedback in the information retrieval community—Ide dec-hi, using term frequency inverted document frequency (TFIDF) weighting on selected collections from the information retrieval community such as CRANFIELD, MEDLINE, and CACM. The evaluations with our hybrid model with these testbeds show that this approach

H. Nguyen (✉)
Department of Mathematical and Computer Sciences, University of Wisconsin-Whitewater,
800 W. Main Street, Whitewater, WI 53190, USA
e-mail: nguyenh@uww.edu

E. Santos Jr.
Dartmouth College Thayer School of Engineering, 8000 Cummings Hall, Hanover, NH
03755, USA
e-mail: eugene.santos.jr@dartmouth.edu

G. A. Tsihrintzis et al. (eds.), *Multimedia Services in Intelligent Environments*,
Smart Innovation, Systems and Technologies 24, DOI: 10.1007/978-3-319-00372-6_3,
© Springer International Publishing Switzerland 2013

retrieves more relevant documents in the first 15 returned documents than the TFIDF approach for all three collections, as well as more relevant documents on MEDLINE and CRANFIELD in both initial and feedback runs, while being competitive with the Ide dec-hi approach in the feedback runs for the CACM collection. We also demonstrate the use of our user model to dynamically create a common knowledge base from the users' queries and relevant snippets using the APEX 07 data set.

1 Introduction

During the last two decades, as a result of the overwhelming number of online and offline information resources, we have witnessed an increasing trend towards development of personalized applications to improve the quality of the users' work. The problem of modeling a user for information retrieval (IR) was identified as early as the late 1970s and continues to be a challenge (examples include [1, 8, 9, 14, 27, 51, 52, 63, 70, 68]. This addresses the lack in accounting for users and their interests within the traditional IR framework. The hypothesis is that by modeling a user's needs, better retrieval of more documents relevant to an individual user can be achieved. The main challenges of employing a user model for information retrieval arise from the fundamental questions of (i) how best to model a user so that it helps an IR system adapt to the user most effectively and efficiently; and, (ii) how best to justify the need of having a user model on top of an IR system from both the perspective of designers of an IR system and that of users of an IR system. It requires a mix of several research areas and approaches to answer these questions, such as artificial intelligence (AI) (natural language processing, knowledge representation, and machine learning), human factors (HF), human computer interaction (HCI), as well as information retrieval. The traditional framework of information retrieval supports very little user's involvement and only considers a user's query with simple or little relevant feedback. On the other hand, the user modeling (UM) community has not taken advantage of the information retrieval techniques except for some traditional techniques such as vector space model and term frequency inverted document frequency (TFIDF) weighting.

In this chapter, we present our research towards improving a user's effectiveness in an information retrieval task by developing a *hybrid user model* that captures *a user's intent* dynamically through analyzing behavioral information from retrieved relevant documents, and by merging it with the components of an IR system in a decision theoretic framework. Capturing user intent for information seeking is an important research topic that has recently drawn much attention from both the academic and commercial communities, with focuses on both image and text retrievals (for examples: [2, 3, 15, 31, 35, 38, 54, 69]. Yahoo! used to maintain a web site at http://mindset.research.yahoo.com that returned the retrieved relevant documents sorted by a user's intention. In our approach, the term *hybrid* refers to the methodology of combining the attributes describing a user's intent with the attributes describing an information retrieval system into a decision theoretic framework. The effectiveness typically refers to the number of relevant documents related to a user's problem and the relevant documents refer to the commonality order in which these documents were returned. The behavioral information refers to the overlapped content shared by these documents. Lastly, the components of an IR system include query, similarity measure, collection, relevancy threshold, and indexing scheme.

The novelty of our hybrid user modeling approach is that it bridges the gap between two worlds: the world of the target IR application and the world of the users of an IR application by integrating techniques from both worlds. The goal is that by using the unified decision theoretic framework, we can maximize the effectiveness of a user's retrieval task with regards to his/her current searching goals. This hybrid user model truly compensates for the lack of information about a user in an IR system by capturing user intent and by adapting the IR system to the user. It also compensates for the lack of information about an IR system within current user modeling research by using the insights from the collections to predict the effectiveness of the proposed adaptations. This is a new approach for both the IR and UM communities. This framework contributes to the IR community by using decision theory to evaluate the effectiveness of the retrieval task. This also provides the fundamental framework in modeling individual context, interests and preferences to make the job of capturing the common interests and effective cognitive styles for a group of users feasible.

In what follows, we present our vision and efforts in bridging these two different areas that are crucial for building a successful IR system. We also bring together some of our past results on developing and evaluating a hybrid user model for information retrieval [45, 47] as evidence of how to apply and assess this framework. We comprehensively evaluate our hybrid user model and compare it with the best traditional approach for relevance feedback in the IR community—Ide dec-hi using term frequency inverted document frequency weighting on selected collections from the IR community such as CRANFIELD, MEDLINE, and CACM. The results show that with the hybrid user model, we retrieve more relevant documents in the initial run compared to the Ide dec-hi approach. Our hybrid user model also performs better with the MEDLINE collection compared to our user modeling approach using only a user's intent [58]. We also demonstrate the use of our user model approach to dynamically create a common knowledge base from the users' queries and relevant snippets for eight analysts using the APEX 07 dataset.

The remainder of this chapter is organized as follows: We begin by reviewing key related work with regards to constructing a user model for improving retrieval performance and using decision theory for information retrieval. Next, some background and the description of our hybrid approach are provided. Then we present the description of the evaluation testbeds, procedures, and evaluation results, which are followed by two applications of hybrid user models. Finally, we present our conclusions and future work.

2 Related Work

The novelty of our approach is to construct a model that integrates information about a user and about an IR system in a decision theoretic framework. Therefore, in this section, we present the related methodologies for developing a user model from both the IR and UM communities and existing work on using decision theory for information retrieval.

2.1 Methodologies for Building a User Model for Information Retrieval

In this chapter, we classify the current approaches to building user models for IR into three main groups based on *how* the models are built *conceptually*. These groups are *system-centered, human-centered (or as also known as user-centered)*, and *connections* (referred later here as *hybrid* approaches) [63]. This categorization scheme highlights the differences in terms of the techniques from the IR and UM communities as well as the need of integrating these approaches.

The methods belonging to the system-centered group focus on using IR techniques to create a user model. The IR community has tried to understand a user's information needs in order to improve a user's effectiveness in an information seeking task by a number of ways such as query specification and relevance feedback/query expansion. Query specification helps a user to describe his/her information needs via a graphical user interface (such as [41, 43]) while relevance feedback/query expansion interactively improves a user's query by learning from the relevant and non-relevant documents in a search (e.g., [12, 25, 29, 53, 55, 66]). Recently, researchers from the IR community have also applied genetic algorithms (GA) [18, 40] and support vector machine (SVM) to relevance feedback [22] as well as applied language modeling techniques to traditional vector space model [72]. These system-centered techniques are also referred to as classical IR solutions and personalized relevance feedback approaches as described in a survey by [68]. Unfortunately, the system-centered approaches still do not describe entirely or even adequately a user's behaviors in an information seeking task. Our approach is different in that we determine a user's intent in information seeking to decide which concepts and relations to add to the original query instead of adding the terms based on their weights directly from relevant and non-relevant documents to a user's original query.

The human-centered or user-centered approaches build a user model by exploring a user's cognitive aspects in a user's interactions with an IR system. Some traditional examples include [9, 14, 30, 70]. The majority of these studies focus on directly or indirectly eliciting a user's interactions, preferences, cognitive searching styles, and domain knowledge to improve interfaces as well as human performance in an IR system. Recently, the information retrieval community has begun to focus more on understanding a user's information needs, a user's behaviors in an adaptive information retrieval system (e.g., [36, 74]), and a user's IR tasks and information seeking strategies (e.g., [34]) to improve user satisfaction with an IR system. Some useful user studies have been conducted to understand more about users' characteristics and task-relevant behaviors such as the study conducted by [11] investigated search intermediaries and search behaviors and the work by [42] analyzed query submissions and navigation using a query log and approximated the user's behaviors to Zipf's Law distribution. Even though the findings of these studies shed some light on the possible directions for improving a traditional IR system, there are still many open questions. One criticism often raised by designers of IR systems is that any model that contains all or even a subset of a user's cognitive dimensions is highly complex and impractical. Furthermore, one key problem that arises with the approaches in this group is that they are concerned primarily with a user's behaviors and have little to say about *why* a person might engage in one particular behavior. In order to find out *why*, we have to establish the relationships between the behaviors and the goals that a user is trying to achieve. What differentiates our approach from the approaches in this group is that we actively establish the relationship between a user's searching behaviors and a user's goals in our hybrid model.

Even though there is still very little overlap between system-centered and user-centered approaches, as identified earlier by Saracevic et al. [63], many researchers have attempted to bridge this gap. We refer to the techniques in this category as *hybrid* approaches. There are two main views of the hybrid approaches: (i) connecting two (or more) different system-centered techniques to construct a model, e.g., applying decision theory techniques to collaborative filtering [46], mixing different models that represent different task-related factors (e.g., [23]) or temporal related factors [10], integrating collaborative filtering techniques with knowledge-based techniques [73] or with content-based techniques [6]; and, (ii) connecting system-centered and user-centered techniques in order to construct a model. Our approach belongs to the latter view. This direction is very challenging because finding a common ground from the different representation schemes and evaluation criteria among different disciplines (such as HF, IR, and AI) is difficult. Some historical examples can be found in the work presented in Logan et al. [39] and Saracevic [62]. In the study described in Logan et al. [39], Galliers' theory of agent communications is applied on the MONSTRAT model [7]. The STRATIFIED model proposed by Saracevic [62] which inspired our approach in this chapter, attempted to resolve the weaknesses of both the system-centered and human-centered approaches. In the STRATIFIED model, both the user and the system sides are viewed as several levels of *strata*. Any level of the user's strata is allowed to interact with any level of the system's strata. The STRATIFIED model is constructed based on the assumption that the interactions between the user and the target IR system help the user's information seeking tasks. Within the last decade, researchers also have explored a user's searching behaviors for constructing a user model. These studies have shown that by understanding a user's searching behaviors, we develop a more flexible IR system with personalized responses to an individual's needs [18, 19, 21, 58, 67]; (Ruthven et al. 2003). For example, in [18, 56], temporal factor, uncertainty, and partial assessment are combined to modify the weight of a term in a relevance feedback process. The main difference between the existing approaches which incorporate a user's searching behaviors discussed above with our approach is that they use a user's search behaviors to modify the *weight of an individual term* while ours uses the captured user intent to modify the *relationships among terms* in a query. A book edited by Amanda Spink and Charles Cole [65] has provided an excellent overview and new research directions to find a central ground for user-centered and system-centered approach in information retrieval. Our work here certainly contributes to this stream of research.

2.2 Decision Theory for Information Retrieval

Even though there is some prior work from both the IR and UM communities that make use of decision theory (e.g., [5, 10, 20]), a decision theoretic framework that fully integrates attributes from both user and system sides has not yet been explored. From the UM community, Balabanovic [5] has used decision theory to elicit a user's preferences over *a set of documents*. Our approach is different from his approach in that, we use decision theory to elicit a user's preferences over *a set of attributes describing a user's current search*. Brown [16] with his *CIaA* architecture has used decision theory to determine if a user model has done a good job in assisting a user based on a set of functions measuring a user's workload, temporal, and physical efforts for a specific task. In our approach, we use the information of the components of an IR system to predict the effectiveness of a user in a current search. From the IR community, researchers in probabilistic IR have used decision theory to determine indexing schemes [20], however, no attention has been paid to users.

3 Capturing a User's Intent in an Information Seeking Task

3.1 Overview

In this work, we develop a hybrid user model to improve effectiveness of an IR system. User models are needed on top of an IR system because the traditional IR framework does not utilize much input from a user except a user's query and some relevance feedback. Without a user model, it is very difficult to determine and update a user's needs. For instance, a user is searching for *"sorting algorithms"* and he possesses knowledge on *"distributed computing"* with an emphasis on *"parallel algorithms"*. He prefers to retrieve papers on specific algorithms rather than survey papers. He also prefers to retrieve as many potentially relevant documents as possible. For this user, a good IR system would display documents about parallel sorting algorithms such as *"odd–even sort"* or *"shear sort"* well before sequential sorting algorithms such as *"bubble sort"* or *"quick sort"*. In other words, a good IR system would pro-actively modify the original request of *"sorting algorithms"* to a request on *"parallel sorting algorithms"* which connects the user's preferences, interests, and knowledge with his current request. Additionally, in order to meet his preference to see as many potentially relevant documents as possible, the threshold for filtering irrelevant documents should be set low to allow those documents that did not contain the exact same terms but may contain synonyms or similar terms to be presented to the user.

Our goal is to improve the effectiveness of a user engaged in an information seeking task by building a user model that integrates information about an IR system and a user in a decision theoretic framework. The components of a typical IR system include *query, indexing scheme, similarity measure, threshold,* and *collection* [4]. Query represents a user's request. Indexing schemes contain domain knowledge represented in hierarchical relations of terms. Similarity measures are a function which determines how similar a user's query and a document from the searched collection is. Threshold is a real number which indicates how we should filter out irrelevant documents. A collection usually consists of a set of documents in a specific topic such as computer science or medicine. Usually, these components are determined when the system is developed and used. Therefore, in order to build our hybrid model, our job now is to determine information about a user and then integrate it with the components of an IR system. Our approach is to capture *user intent* in an information seeking task. We partition it into three formative components: Interests accounts for *what* a user is doing, Preferences captures *how* the user might do it, and Context infers *why* the user is doing it. The rest of this section presents the process of capturing user intent in an information seeking task [58, 59, 61]. In the next section, we combine the user's intent with the attributes of an IR system to create our hybrid user model.

We capture the Context, the Interests, and the Preferences aspects of a user's intent with a context network (*C*), an interest set (*I*), and a preference network (*P*). Before we describe the representation of each component, we go over how documents are represented because this representation is also used in our Context network. Each document in the target database is represented as a document graph *(DG),* which is a directed acyclic graph (DAG) containing two kinds of nodes: concept nodes and relation nodes. Concept nodes are noun phrases such as *"slipstream"* or *"comparative span"*. We capture two types of relations: set-subset (denoted as *"isa"*) and related to (denoted as *"related_to"*). A relation node of a DG should have concept nodes as its parent and its child. An example of a part of a *DG* generated from a document in the CRANFIELD collection is shown in Fig. 1. The highlighted terms in the Fig. 1a are concept nodes in the Fig. 1b. We developed a system to automatically extract *DG* from the text [75]. We extracted noun phrases (NPs) from text using Link Parser [64]; these NPs will become

Experimental investigation of the aerodynamics of a wing in a slipstream. An experimental study of a wing in a propeller slipstream was made in order to determine the spanwise distribution of the lift increase due to slipstream at different angles of attack of the wing and at different free strea m to slipstream velocity ratios. The results were intended in part as an evaluation basis for different theoretical treatments of this problem. The comparative span loading curves, together with supporting evidence, showed that a substantial part of the lift increment produced by the slipstream was due to a /destalling/ or boundary-layer-control effect. The integrated remaining lift increment, after subtracting this destalling lift, was found to agree well with a potential flow theory. An empirical evaluation of the destalling effects was made for the specific configuration of the experiment

(a)

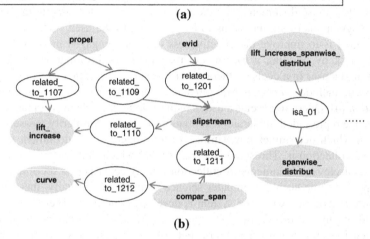

(b)

Fig. 1 a An example of a document from the CRANFIELD collection. **b** A small part of the document graph generated from this document

concept nodes in a *DG*. The relation nodes are created by using three heuristic rules: *noun phrase heuristic*, *noun phrase-preposition phrase heuristic*, and *sentence heuristic* (see [58] for specific details). Each query is also represented as a query graph *(QG)* which is similar in representation to the document graph.

3.2 Interest Set

The *Interest set* represents *what* a user is currently focusing on. Each element of this set describes an interest concept a and its associated interest level $L(a)$. Interest concept refers to the terms that the user is interested in and interest level is a real value in the range of [0,1] that represents the user's emphasis on a particular concept. We use the intersection of *DGs* of retrieved relevant documents to create and update an interest set. The algorithm of finding the intersections of retrieved relevant documents is shown in Fig. 2. Each concept in this intersection will be added to the current interest set with the interest level being the ratio of frequency

```
/* D is the set of m retrieved relevant document graphs */
Vector Intersection(D) {
        Vector J= Ø
        For i = 1 to m{
                For each node c in Di do
                        frequency(c)=0
                        If c is a concept node then
                                For j = 1 to m
                                If i does not equal to j and (Dj contains c) then
                                        frequency (c) ++
                                        if (frequency(c) > threshold) then J = J + c
        }
        For i=1 to m do {
                For each node r in Di do
                        frequency(r) = 0
                        For k=1 to m do
                                If (r is a relation node and (its parent and its child are in J and
                                k does not equal to i) then frequency(r) ++
                                If (frequency(r) > threshold ) then J = J+ (r's parent - r - r's child)
        }
        return J
}
```

Fig. 2 Pseudo code of algorithm to find intersections of retrieved relevant documents

of a specific concept node over the total concept nodes in the intersection and if this interest level is greater than the user-defined threshold for Interest list.

A user's interests change over time, therefore, we re-compute the interest level $L(a)$ for each concept a in the existing interest set after each query as follows:

$$L(a) = 0.5\left(L(a) + \frac{p}{q}\right) \tag{1}$$

in which p as the number of retrieved relevant documents containing a and q as the number of retrieved documents containing a. If $L(a)$ falls below a user-defined threshold value, the corresponding interest concept a is removed from the interest set.

3.3 Context Network

The Context network captures a user's knowledge in a specific domain, which is represented in terms of concepts and relationships among these concepts. The basic structure of the Context network is similar to the representation of a DG. We choose the DG representation for representing a Context network because it satisfies two requirements: (i) it can be generated dynamically, and (ii) it represents concepts and relationships among concepts visually. In a Context network, each node also has a weight, value, and bias. These attributes are used in our re-ranking algorithm to infer the current interests. The weight of a node represents its importance assigned initially by the system. The concept nodes and *"isa"* relation nodes have initial weights equal to 1 while the *"related to"* relation nodes have weight equal to 0.8 in this implementation. The value of a node represents its importance to a user and is a real number from 0 to 1. The bias of a node represents whether this node is actually in the user's interests or not. Each node's weight, value and bias are used by a spreading activation propagation algorithm. The basic principle of this algorithm is that a node located far from an interest concept, which has been indicated as being interested by the user, will be of less interest to the user. The spreading activation propagation algorithm [58] is shown in Fig. 3.

Note that in this algorithm, we treat the nodes with one parent differently from those with multiple parents. The intuition is that a single parent only will have stronger influence on its children while the influence from multiple parents needs to be aggregated to avoid bias from a specific parent.

A Context network is dynamically constructed by finding the intersection of all document graphs representing retrieved relevant documents. The algorithm to find the intersection of the retrieved relevant document is shown earlier in the previous subsection. We use the set of common sub-graphs of the retrieved relevant documents in the intersection. If a sub-graph is not currently in the Context network, it is added to the Context network accordingly. If the addition results in a loop, it will be skipped. We do not consider loops because we want to maintain the rationality of the relationships among concepts and also ensure that our spreading activation inference algorithm on a Context network terminates properly. This does not cause any problems because of the semantics of the relationship that we capture. Specifically, a loop cannot occur for *"isa"* relationships. If a loop occurs for *"related_to"* relationships, this is already covered by the transitive and reflexive attributes of this type of relationship so the extra link is unnecessary.

A new link between two existing concepts in a context network will also be created if two concepts are indirectly linked in the set of common sub-graphs and the frequency of these links exceeds a certain user-defined threshold. An example of a Context network of a user model from one of the experiments conducted in this chapter is shown in Fig. 4.

3.4 Preference Network

The *Preference network* represents *how* a query is formed [58]. We chose Bayesian networks [32] to represent a preference network because they offer visual expressiveness, and the ability to model uncertainty. A user shows his/her preferences in the ways to modify a query by choosing from a class of tools. We currently choose two tools to implement in our Preference network which are filters and expanders. A filter is a tool that searches for documents that narrows the search topics semantically and an expander that searches for documents that broaden the search topics semantically. We chose these two tools because they are

```
Vector SpreadingActivation(I, Q, C) {
    /*Initializing bias is executed as follows: (i) set the bias equal to 1 for every concept
    found both in the current context network and in the current query graph; and (ii) set bias
    equal to the interest level for every interest concept found both in the current context
    network and in the current interest set. */
    initBias();

    /* Then we compute and sort all the nodes in the concept network C by its depth.
        d(a) = 0 if the node has no parents
    d(a) = max(d(p(a))) + 1 with p(a) is a parent of a.
    */
    Vector nodesInContextNetwork = sortBasedOnDepth(C);
    For each node a in C do {
        If (a.getNumberOfParents() == 0) then {
            sum = 0
```

if (bias > sum) then $value = \dfrac{(sum + bias)}{2}$

else $value = 0$

```
        }

        If (a.getNumberOfParents() == 1) then {
```

sum = value(p(a))* weight(p(a))

if (bias > sum) then $value = \dfrac{(sum + bias)}{2}$

else $value = sum$

```
        }
        If (a.getNumberOfParents() > 1) then {
```

$$sum = \dfrac{1}{1 + e^{-\dfrac{\sum value(p_i(a))*weight(p_i(a))}{n}}}$$

/* $p_i(a)$ is a parent of a node a, n is the total number of all parent nodes. We
chose this function to ensure that the value of each node is converged to 1 as the
values and weights of its parents are increasing */

if (bias> sum) then $value = \dfrac{(sum + bias)}{2}$

else value = sum

```
        }
    }
    Vector newInterests = ∅
    Vector conceptList = sortConceptNodesByValues(C);
    For each c in conceptList do {
        If (c.getValue() >= short term interest threshold) then
            newIntersts=newInterests + c
    }
    Return newIntersts
}
```

Fig. 3 Pseudo code of our spreading activation algorithm

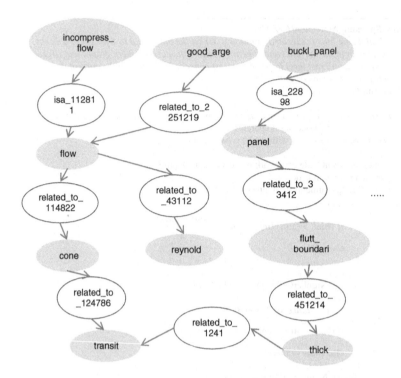

Fig. 4 A part of a user's context network

Fig. 5 Conceptual structure
of a preference network

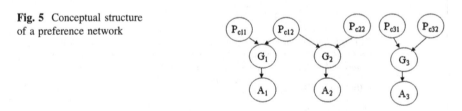

similar to two typical methods of refining a search query in information seeking:
specification and generalization [37].

There are three kinds of nodes in a preference network: pre-condition node P_c,
goal node G, and action node A, as shown in Fig. 5. A P_c represents the
requirements of a tool, such as the query and/or the concepts contained in the
current interest relevancy set. If a P_c represents a concept from a user's interest set,
its prior probability will be set as its interest level. If a P_c represents a query, its
prior probability will be set as its frequency. A goal G represents a tool, which is
again a filter or an expander in the current design. The conditional probability table
of each goal node is similar to the truth table of logical AND. Action node
A represents an action associated with each goal node. Each goal node is associated with only one action node. The conditional probability of the action node will
be set to 1 if the corresponding tool is chosen and to 0, otherwise.

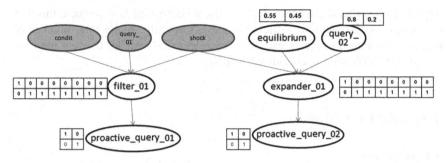

Fig. 6 Example of a preference network

To manage the size of the conditional probability table of each goal node when the interest set grows, we apply a divorcing technique [50] through noisy-AND nodes. An example of a preference network is shown in Fig. 6. There are five pre-condition nodes in this network in which three pre-condition nodes represent the user's interests (labeled as *"condit"*, *"shock"*, and *"equilibrium"*) and two nodes represent the user's queries (labeled as *"query_01"* and *"query_02"*). The high-lighted nodes in this figure are the ones set as evidence. The interest concept *"equilibrium"* has interest level as 0.55 and therefore its prior probability is set as (P(*equilibrium* = true) = 0.55 and P(*equilibrium* = false) = 0.45) as shown in Fig. 6. Similarly, the prior probability of the pre-condition node query_02 is set as P(*query_02* = true) = 0.8 and P(*query_02* = false) = 0.2. There are two goal nodes and two actions in this preference network. Figure 6 also shows the con-ditional probability tables associated with the goal nodes *filter_01* and *expan-der_01*, which are similar to the logical AND truth table. Each entry in a conditional probability table of a preference network shows the conditional prob-ability of each outcome of the goal node given outcomes of all pre-condition nodes (e.g., P(*filter_01* = true|*query_01* = true, *condit* = true, *shock* = true) = 1). A conditional probability table associated with an action node shows the proba-bility of each outcome of the action given the outcome of each goal (e.g., P(*proactive_query_01* = true|*filter_01* = true) = 1).

We use a user's query and retrieved relevant documents to create and update a preference network. If this query or some parts of it have been encountered before, the existing pre-condition nodes representing previously asked queries in the preference network that match the current query will be set as evidence. Each interest concept from the current interest set is added to the preference network as a pre-condition node and set as evidence. If the user's query is totally new and the preference network is empty, the tool being used by the user is set to the default value (a filter) and a goal node representing the corresponding tool is added to the preference network. Otherwise, it is set to the tool being represented by the goal node with highest marginal probability. Each action node represents a way to construct a modified query based on the current tool, interests and user query. A new tool is chosen based on the probability that a new network with this tool will improve the user's effectiveness. Specifically, we determine how frequent this tool has helped in the previous retrieval process. Currently, if the total number of retrieved relevant documents exceeds a user-defined threshold, the tool used for the query modification is considered helpful.

4 Hybrid User Model

4.1 Overview

We combine the user intent with the elements of an IR application in a decision theoretic framework to construct a hybrid model to improve a user's effectiveness in a search. Our solution is to convert this problem into a multi-attribute decision problem and use multi-attribute utility theory [33] in which a set of attributes is constructed by combining the set of attributes describing a user's intent and the set of attributes describing an IR system. In a multi-attribute utility model, the decision is made based on the evaluation of *outcomes* that are generated by the actions performed by a user. In this problem, the outcome space represents the set of *all* possible combinations of a user's intent and components of an IR system. Each outcome represents a specific combination of a specific user intent and specific collection, indexing scheme, query, similarity measure, and threshold. There are two reasons for using this multi-attribute utility model. First, the estimation of a user's effectiveness in a search with respect to the searching goal is a problem of preference elicitation. It represents a user's preferences over a space of possible sets of values describing a user and describing an IR system. Second, the framework of a multi-attribute decision problem allows us to use the elicitation techniques from the decision theory community to decide which combination will be likely to produce the most effective search. We use the literature on predicting query performance in IR [28] and computing dissemination thresholds in information filtering (IF) [13] in our elicitation technique.

In this hybrid user model, eight attributes are initially considered from the set of attributes describing user intent and the set of attributes describing an IR system. They are: I (a user's Interest), P (a user's Preferences), C (a user's Context), In (Indexing scheme), S (similarity measure), T (threshold), D (Document collection) and Q (a user's query). From this list of eight attributes, we know that I, P and C have been used effectively to modify a user's query Q [58, 60]. Therefore, the attribute Q can subsume the attributes I, P and C. In the traditional IR framework, the indexing scheme In, document collection D, and similarity measure S are decided when designing a system and shall remain unchanged during a search process. These three attributes do not participate in the decision making process.

After reducing the number of attributes to core attributes, we focus on only two attributes Q and T. Note that this is not a simple integration between query and threshold attributes because we model the non-deterministic behavior of a user in information seeking through a user's interests, preferences, and context and use this information to modify a user's query. Therefore, by using these two attributes, we reflect well both a user and a typical IR system and reduce the complexity of the problem. We then evaluate each outcome through a real value function. We also make an assumption that these two attributes are preferentially independent [33]. Thus, this value function representing a user's preferences over these two attributes can be defined as follows:

$$V(Q, T) = \lambda_1 V_1(Q) + \lambda_2 V_2(T) \tag{2}$$

where λ_i represents the importance of attribute i to the user, and V_i is a sub-value function for the attribute i with $i = 1$ or $i = 2$. This value function is generic for all IR systems and all type of users.

In this hybrid user model, we do not work directly with the value functions because it is very difficult to elicit the coefficients λ_i. Instead, we determine the partial value function which consists of two sub-value functions: one over query and one over threshold. The partial value function implies that an outcome x_1 with the value (x_{11}, x_{12}) is preferred to an outcome x_2 with value (x_{21}, x_{22}) if and only if

$$x_{li} \geq x_{2i} \text{ for all } i = 1, 2, \text{ and}$$
$$x_{li} > x_{2i} \text{ for some } i.$$

For each sub-value function, each attribute needs to be evaluated with respect to a user's effectiveness in achieving a searching goal at a given time. We assume that a user's searching goal at any given time is to retrieve many relevant documents *earlier* in the information seeking process. Therefore, we choose the average precision at three point fixed recalls as the effectiveness function because it measures both the percentage of retrieved relevant documents and the speed of retrieving these documents. This function is computed by calculating the precision values at various recall points. For instance, we calculate the average precision (*ap*) at three point fixed recalls of 0.25, 0.5, and 0.75 for a query Q by:

$$\mathbf{ap} = \frac{P_{0.25} + P_{0.5} + P_{0.75}}{3} \tag{3}$$

where P_{recall} is computed as follows:

$$P_{recall} = \frac{m_{recall}}{M} \tag{4}$$

where $m_{recall} = recall*N$ with N is the total relevant documents for this query Q, and M is the number of the retrieved documents when m_{recall} relevant documents are retrieved.

4.2 Sub-Value Function Over Query

We take advantage of the research on predicting query performance in the IR community to construct a sub-value function over a query. Basically, we have chosen the standard deviation of a query's terms' *inverted document frequency* (*idf*) as the *core* of this sub-value function. The main idea of the *idf* measure is that the less frequent terms in a collection are the terms with more discriminating power. The primary reasons for our choice are (i) the standard deviation of *idf* of a query's terms (also known as the distribution of informative amount in query terms [20] has shown relatively good positive correlation with the average precision metric, and (ii) it can be computed in a pre-retrieval process.

Recall that each query in this approach is represented by a query graph which contains relation and concept nodes [58, 60]. Therefore, we created the sub-value functions for the concept nodes and for the relations. A sub-value function for the concept nodes is computed as follows:

$$V_c(Q) = \sigma_{idf-c}(Q) \tag{5}$$

in which

$$\sigma_{idf-c}(Q) = \sqrt{\frac{1}{n}\sum\nolimits_{c\in Q}(idf_c(c) - \mu_{idf-c}(Q))^2} \tag{6}$$

with n as the number of concepts in Q

$$\mu_{idf-c}(Q) = \sum\nolimits_{c\in Q}\frac{idf_c(c)}{n} \tag{7}$$

and,

$$ldf_c(c) = \frac{log_2(N + 0.5)/N_c}{log_2(N + 1)} \tag{8}$$

where N is the total number of documents in a collection and N_c is the total number of documents containing the concept c.

Similar to the sub-value function computed based on information about concept nodes, we define the sub-value function computed from information about the relation nodes. A relation r in Q is represented as a tuple (c_1, r, c_2) in which c_1 and c_2 are two concept nodes, and r is either *"isa"* or *"related to"* relation:

$$V_r(Q) = \sigma_{idf} - r(Q) \tag{9}$$

in which

$$\sigma_{idf-r}(Q) = \sqrt{\frac{1}{n}\sum\nolimits_{r\in Q}(idf_r(r) - \mu_{idf-r}(Q))^2} \tag{10}$$

with n is the number of relation r in Q

$$\mu_{idf-r}(Q) = \sum_{r \in Q} \frac{idf_r(r)}{n} \qquad (11)$$

and,

$$idf_r(r) = \frac{log_2(N + 0.5)/N_r}{log_2(N + 1)} \qquad (12)$$

where N is the total number of documents in a collection and N_r is the total number of documents containing the relation r.

4.3 Sub-Value Function for Threshold

We take advantage of research from adaptive thresholding in information filtering, specifically the work in [13], to construct a sub-value function for thresholds. We choose the threshold of the last document seen by a user and the percentage of returned documents preferred to be seen by a user as the *core* of our sub-value function.

For each query, the initial threshold can be determined as:

$$T_0 = p^* N_0 \qquad (13)$$

where N_0 is the number of documents returned at time 0, and p is the percentage of retrieved documents that a user wants to see, e.g., highest 10 %, highest 20 % or highest 80 % of retrieved documents. For the first time a user is using the system, this number is elicited by directly asking the user. If this is not the first time, then p is determined as follows:

$$p = \frac{l}{L} \qquad (14)$$

where l is the number of documents that are returned in the previous retrieval and seen by the user and L is the number of documents that contain at least one concept in the query of the previous retrieval.

The threshold is updated by using the approach reported in [13]:

$$T_{(t+1)} = T_t + \frac{sim(d_{last}) - T_t}{e^{\frac{(R_t - \lambda)}{\Phi}}} \qquad (15)$$

where $\lambda = 1300$ and $\varphi = 500$ and R_t is the total number of relevant documents at time t, and d_{last} is the similarity of the last retrieved document in the previous retrieval. The values of these λ and φ constants are obtained from experimental results in [13]. The logic for this approach is that if the number of retrieved relevant documents is small, and the difference between the similarity of the last returned documents and the threshold is big, then we need to decrease the threshold considerably in order to retrieve more relevant documents. Otherwise, we may not need to decrease the threshold.

This method of updating thresholds is chosen because it is light-weight and can be computed in the pre-retrieval process. It also has been shown to correlate well with average precision in [13].

The sub-value function for the threshold attribute will then be defined as follows:

$$V(T) = \begin{cases} 1 \text{ if } T > T_t \\ 0 \text{ otherwise} \end{cases} \tag{16}$$

4.4 Complexity of Hybrid User Model

The process of computing $idf_c(c)$ for every concept and every relation can be done offline. The complexity of this process is $O(nm)$ with n being the number of documents and m being the maximum number of nodes in a document graph. The only online algorithms are the computation of $V_c(Q)$ and $V_r(Q)$ for those concepts and relations included in a user's query. The computation of $V_c(Q)$ has complexity $O(l_c log_2(N) + l_c)$ with l_c being the number of concepts in a query and N being the number of concepts in the collection. Similarly, the computation of $V_r(Q)$ has complexity $O(l_r log_2(N) + l_r)$ with l_r being the number of relations in a query, and N being the number of relations in the collection.

4.4.1 Implementation

We embed and use this hybrid user model in an IR system as follows:

- A user logs into an IR system. If the user is new, then he/she is asked for his/her preferred percentage of documents needed to be returned p.
- The user issues a query Q. The user's query is modified using the information contained in the Interest, Preference and Context (the pseudo code is shown in Fig. 7). Assuming that there are m goals fired in the Preference network, each goal generates a query, so we have the query sets $\{Q_1, Q_2, ..., Q_m\}$.
- Use the sub-value function to evaluate each Q_i. Choose the query with the highest sub-value function evaluation. Determine T_0 for initial threshold.
- Send the query with the highest value evaluated by the sub-value function to the search module, perform the search, filter out the documents based on the value of the threshold, and display the results to the user.
- We update the sub-value function $V(T)$. If a new query is issued, re-compute the threshold depending on the number of documents seen in the previous step.

ModifyQuery (I,P,C,Q) {

- Using the spreading activation algorithm described earlier to reason about the new set of interest *I'*.
- Set as evidence all concepts of the interest set *I'* found in *P*.
- Finding a pre-condition node representing a query in *P* which has associated *QG* that completely or partially matches against *q*. Set it as evidence if found.
- G = Performing belief updating on the preference network P. Choose top *m* goal nodes from preference network with highest marginal probability values.
- For every goal node in *G* do
    - If the query has been asked before and the user has used this goal, replace the original query subgraph with the graph associated with the action node of this goal.
    - If the query has not been asked before and the goal node represent a filter:
    For every concept no de q_i in q, we search for its corresponding node cq_i in the context network *C*. For every concept *i* in *I'*, we search for its corresponding node c_{ii} in the context network such that c_{ii} is an ancestor of cq_i. If such c_{ii} and cq_i are found, we add the paths from context network between these two nodes to the modified query graph.
    - If the query has not been asked before and the goal node represents an expander:
    For every concept node q_i in the user's query graph *q*, we search for its corresponding node cq_i in the context network *C*. For every concept *i* in *I'*, we search for its corresponding node c_{ii} in the context network such that c_{ii} is a progeny of cq_i. If such c_{ii} and cq_i are found, we add the paths from context network between these two nodes to the modified query graph.
    - Add this modified query graph Q_i to the query set

}

Fig. 7 Pseudo code for modifying a user's query using captured intent

An example of an original query and a proactive query is shown in Fig. 8a, b respectively. The shaded nodes in Fig. 8b are the nodes added as a result of the algorithm in Fig. 7.

5 Evaluation

5.1 Objectives

The objectives of this evaluation are as follows:

- To assess whether the hybrid user model improves user effectiveness in an information seeking task. The intuition behind this objective is that we want to verify if our user model helps users retrieve *more* relevant documents *earlier* in an information seeking process.

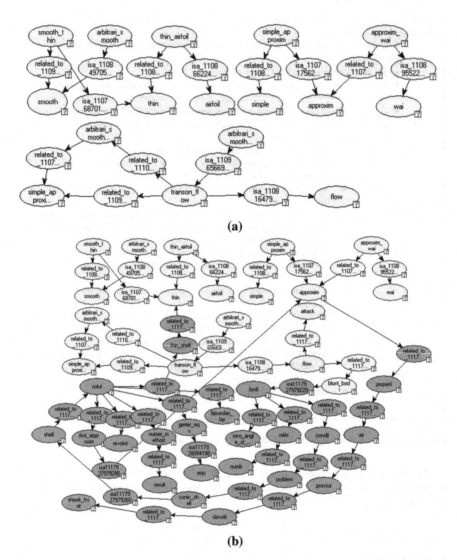

Fig. 8 **a** An example of an original query **b** the corresponding proactive query

- To compare our hybrid user model with the existing approaches from the IR community by using collections, metrics and procedures from the IR community. We set this objective to see where our approach stands and what we can do to make it work better.
- To assess the effect of having a long-term user model with an IR system. User models offer a way to re-use knowledge gained in the past to speed up and enhance the information seeking task at present. With this objective, we hope to understand more about the effectiveness of this reusability in combination with the change of a user's current focuses.

5.2 Testbeds

We decide to use three small collections, CRANFIELD, CACM, and MEDLINE, because a common testbed for evaluating an adaptive IR system still does not exist [71]. Also, these three collections have been extensively studied in our previous work [48, 49, 58]. Even though these collections are not made for evaluating user modeling techniques, their sizes are appropriate and they have been used in the IR community to evaluate the effectiveness of relevance feedback techniques [22, 40, 57]. The CRANFIELD collection contains 1400 documents and 225 queries on aerodynamics; CACM contains 3204 documents and 64 queries in computer science and engineering (CSE); while MEDLINE contains 1033 documents and 30 queries in the medical domain [57]. We use the complete set of queries from these collections in our evaluation.

5.3 Vector Space Model and Ide dec-hi

We compare this user modeling approach against the vector space model using term frequency inverted document frequency (TFIDF) weighting scheme and the Ide dec-hi technique for relevance feedback [57]. TFIDF and the Ide dec-hi techniques are very well-documented [26, 40, 57] and thus make it easier and more reliable for re-implementation. Secondly, the Ide dec-hi approach is still widely considered to be the best traditional approach in the IR community [26, 40, 57]. It offers an opportunity to see where we stand with other approaches as well.

The main idea of the Ide dec-hi is to merge the relevant document vectors into the original query vector. This technique automatically re-weights the original weight for each term in the query vector by adding its corresponding weights from the relevant documents directly and subtracting its corresponding weight from the first non-relevant document. For the terms which are not from the original query vector but appear in the relevant documents, they are added automatically to the original query vector with their associated weights. For the terms which are not from the original query vector but appear both in the relevant document and non-relevant documents, their weight would be the difference between the total weights of all relevant documents and the weight in the first non-relevant document. For the terms which are not from the original query vector but appear only in non-relevant documents, they are not added to the original queries vector with negative weights [26].

The formula for Ide dec-hi is:

$$Q_{new} = Q_{old} + \sum D_i - D_j \tag{17}$$

in which Q_{new} and Q_{old} represent the weighting vector for the modified query and the original query, respectively; D_i represents the weighting vector for any relevant document and D_j represents the weighting vector for the first non-relevant document.

5.4 Procedures

We initially followed the traditional procedure for evaluating any relevance feedback technique as described in [57]. However, this procedure did not provide a way to assess the new features of our hybrid model. Thus, we employ a new evaluation procedure to assess the use of knowledge learned over time.

5.5 Traditional Procedure

We first apply the traditional procedure used in [57] for both Ide dec-hi/TFIDF and the IR application enhanced by our hybrid model. We issue each query in the testbed. We then identify the relevant and irrelevant documents from the first 15 returned documents, and use them to modify the query proactively. For the Ide dec-hi/TFIDF, the weight of each word in the original query is re-computed using its weights in relevant documents and the first irrelevant document. The words with the highest weights from relevant documents are also added to the original query. For our user modeling approach, we start with an empty user model and add the concept and relation nodes to the original query graph based on the procedure described in previous sections. The structure of a query graph is similar to the structure of a document graph and we construct it from a user's query. We choose to use the sub-value function $V_c(Q) = \sigma_{idf-c}(Q)$ over concept nodes in a query as a sub-value function for the query because it is simple and easy to implement. We then run each system again with the modified query. We call the first run, the *initial run* and the second run, the *feedback run*. For each query, we compute average precision at three point fixed recalls.

5.6 Procedure to Assess Long-Term Effect

In this procedure, we would like to assess the effect of knowledge learned from a query or a group of queries. We start with an empty user model and follow similar steps as described in the traditional procedure above. However, we update the initial user model based on relevance feedback and we do not reset our user model, unlike the traditional procedure above.

6 Results and Discussion

6.1 Results of Traditional Procedure

The average precision at three point fixed recall of the initial run and feedback run using original collection of the experiments in standard procedure for CRAN-FIELD, CACM and MEDLINE is reported in Fig. 9. Also in this figure, we report the results for TFIDF/Ide dec-hi approach. In the traditional procedure, it shows that using CRANFIELD collection, we achieve better results in both initial run and feedback run compared to TFIDF/Ide dec-hi approach. For the MEDLINE collection, we clearly achieve better results in the initial run compared to TFIDF approach. Lastly, we achieve competitive performance using CACM collections compared to Ide dec-hi with TFIDF.

Compared to our user model that contains only a user's intent on the entire CACM and MEDLINE collections (that model is referred to as *IPC* model in [58]), as shown in Fig. 10, we achieve only competitive results in both runs for the CACM collection while we are clearly better in the initial run and competitive for feedback run for the MEDLINE collection. In the earlier evaluation done with the *IPC* model, we only selected a small subset of queries from the CRANFIELD collection while we used the entire CRANFIELD collection in this evaluation for the hybrid model. Therefore, no comparisons were made with CRANFIELD.

6.2 Results of New Procedure to Assess Long-Term Effect

The results of our procedure to assess the long-term effects of our hybrid approach are shown in Fig. 11. It shows that by using our hybrid model, the precision of the feedback runs is always higher than those of the initial runs. In the MEDLINE

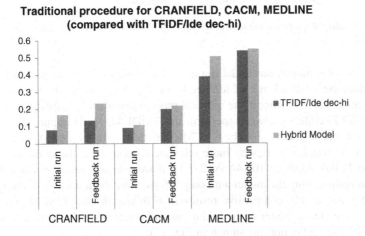

Fig. 9 Result for traditional procedure while comparing with TFIDF/Ide dec-hi

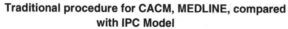

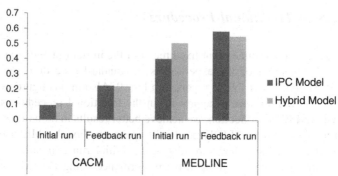

Fig. 10 Result for traditional procedure while comparing with IPC model

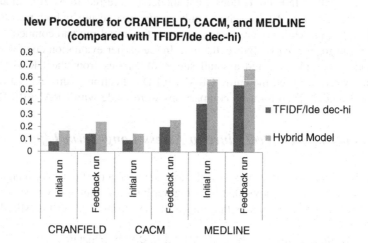

Fig. 11 Results for the procedure to access long-term effect compared with vector space model/ Ide dec-hi

collection, for example, our initial run using knowledge of learned queries is even better than the feedback run of Ide dec-hi/TFIDF. That means the relevant documents are retrieved earlier in the retrieval process than the other approach. For the CRANFIELD collection, we outperform the TFIDF/Ide dec-hi approach in both initial and feedback runs. For the CACM collection, with the new procedure, we maintain the trend of retrieving more relevant documents in the initial run compared to TFIDF approach (0.144 vs. 0.091). If we compare these results with our previous results using the model with only information about a user's intent in [48, 58], we achieve only competitive results in both runs for the CACM collection while we are clearly better in the initial run and competitive for feedback run for the MEDLINE collection (as shown in Fig. 12).

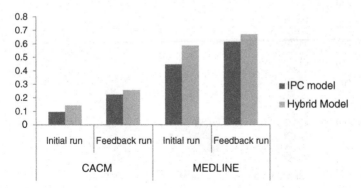

Fig. 12 Results for the procedure to access long-term effect compared with the IPC model

7 Discussion

In summary, with regards to our objectives, we have found the following from this evaluation:

- Effectiveness of the model: Our hybrid user models help retrieve more relevant documents earlier in the initial run compared to both vector space model with TFIDF and *IPC* model.
- Usefulness of having a long term model: In this hybrid model, the long-term and short-term interests of a user in information seeking are combined and balanced. The long-tem interests here refer to knowledge learned from a query or a group of queries while short-term interests refer to knowledge learned from a current query. We clearly perform better than the traditional vector space model with TFIDF/Ide dec-hi on the CRANFIELD and MEDLINE collection and we are competitive on the CACM collection.

As we may have noticed, from the results in both the experiments with traditional procedure and new procedure, our hybrid user model performs better in the initial run but only performs competitively in the feedback run compared to the vector space model using TFIDF weighting scheme and Ide dec-hi approach. The main reason is that our hybrid user model retrieved more relevant documents in the top 15 in the initial run, as shown in Fig. 13. Therefore, as a result, there are less relevant documents left to be retrieved in the feedback run.

In comparison with the *IPC* model, we achieve only competitive results for the CACM but clearly perform better for the MEDLINE. The main reason for this behavior of the hybrid model depends on the distribution of the value function for queries, the distribution of the standard deviation of *idf* for retrieved relevant documents and all relevant documents, and the tools that are chosen to modify each query in the testbed. Further studies on this will be conducted in our future work.

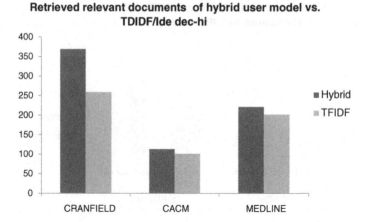

Fig. 13 Retrieved relevant documents in the top 15

8 Application of Hybrid User Model

In this section, we propose a framework in which our hybrid user model can be used to create a common knowledge based on the individuals' knowledge. This demonstrates how the hybrid user model can be applied to develop shared knowledge bases that are used in any applications.

Our hybrid user models can be used to dynamically create the common knowledge bases that contain general domain information as well as provide effective ways of learning a specific domain. Figure 14 describes the pseudo code for creating a general knowledge base that contains the concepts and relationships of concepts from a set of Context networks. The novelty of this approach as compared to static knowledge bases such as Wordnet [44] is that the collective domain knowledge is generated dynamically from a user's interactions in an information seeking process. This approach also differs from the approaches that use traditional vector space model with TFIDF to general domain ontology (such as the work in [24]) in that it provides the relationships between concepts including the concepts that do not occur in the same document as opposed to a list of independent concepts.

We have created a common knowledge base from a set of queries and relevant snippets that were issued by eight intelligent analysts in the APEX 07 data set. This data set was created by the National Institute of Standards and Technology (NIST) to simulate an analytical task in the intelligence community. This collection included eight analysts (namely APEXB, APEXC, APEXE, APEXF, APEXH, APEXK, APEXL, and APEXP), their recorded actions over time, and their final reports. In the APEX data set, there are eight types of actions: "*Start application*",

```
CommonKnowledgeBase(HybridUserModel u[] ) {
    knowledgebase = null;
    For each hybrid user model u_i do {
            Retrieve the context network: c_i = u_i. getContextNetwork();
            Node n_i = c_i. getFirstNode();
            while ( n_i !=null ) {
        if ( knowledgebase does not contain n_i and n_i is a concept node) {Add n_i to
        knowledgebase;
                            Find all of its grand children to see if they are included in
                            the knowledgebase. If they are, add the links
                            Find all of its grant parents to see if they are included in the
                            knowledgebase. If they are, add the links.
                    }
                    n_i = c_i.getNextNode(n_i);
            } // end of while
    } //end of for
    Return knowledgebase;
}
```

Fig. 14 Pseudo code of creating common knowledge base from a set of hybrid model

"*Search*", "*Retain*" (happens when an analyst bookmarks, prints, saves a document, or cuts and pastes information from a document to his/her report), "*Access*" (happens when an analyst views a document), "*Make Hypothesis*", "*Associate Evidence*" (happens when analyst links a document or a snippet to a hypothesis), "*Assess*" (happens when analyst assesses how relevant a document or snippet is to a hypothesis), and "*Discard*" (happens when a user discards evidence). In this experiment, we create a user model for each analyst using only his Retain and Search actions. Figure 15 shows the common knowledge base is created by using the algorithm shown in Fig. 14 for these eight analysts from 12/07/2007 (15:42:44) to 12/11/2007 (14:03:58).

We can also use our hybrid user models to develop a collective strategy by constructing a library of the most effective queries. This is done by combining the set of queries from users and ranking them by the value of the sub-value function over each query. The queries used in the IR application enhanced by this hybrid user model are the ones with highest value functions for each user. Figure 16 shows pseudo code of the algorithm to create a library of queries for a group of users. The query library can be used to improve the retrieval effectiveness of the collaborative information retrieval activities in which concepts from the queries or queries themselves can be recommended to the users of a collaborative group to improve the number of relevant documents retrieved in the information seeking process. For our experiments with CRANFIELD, CACM and MEDLINE collections in this paper, we currently maintain the library with 225, 64 and 30 proactive queries for these three collections, respectively.

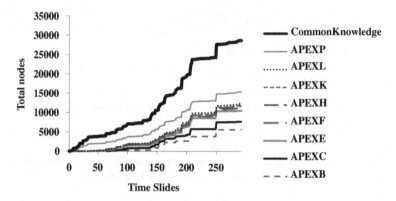

Fig. 15 Common Knowledge base is created using the APEX 07 data set

9 Conclusions and Future Work

Employing a cognitive user model for IR is a critical and challenging problem to help an IR system retrieve more relevant documents more quickly. This model accounts for a user's interests and preferences in information seeking. Even though there are many interesting published research approaches from both the IR and UM communities that try to address this problem, this work stands out because it fills in the gap between two views: system-focus view from IR and the user-focus view from UM. In this summary, we recap the key contribution of this work, and then present different ways of using this framework in a personalized information retrieval for individuals or for a group.

The key contribution is the methodology to construct a hybrid user model. Information about a user and information about an IR system are combined in a decision theoretic framework in order to maximize the effectiveness of a user in an information seeking task with regards to their current searching goals. The hybrid user model truly compensates for the lack of information about a user of an IR system by capturing user intent and adapting the IR system to a user. It also compensates for the lack of information about an IR system by using the information about collections to predict the effectiveness of adaptations. This has never been done before in either the IR or UM communities. By doing so, we allow a user to influence, at a deeper level, the IR system rather than just through the query. Above all, we can use the results of the existing research in IR, IF and UM within a decision theoretic framework to predict the effectiveness of the retrieval task.

The solution presented in this chapter is (i) to convert this problem into a multi-attribute decision problem in which a set of attributes is constructed by compiling the set of attributes describing a user's intent and the set of attributes describing an IR system; and, (ii) to take advantage of research results from the IR community on predicting query performance and from the IF community on determining dissemination threshold to determine sub value functions for these chosen attributes.

```
CommonQueryLibrary (HybridUse rModel u[]) {
        Library = null;
        For each user model u_i do {
                Retrieve a user's current set of queries: q_i = u_i.getQuerySet();
                Sort this set by values of value functions over query
                For each query in this set q_{ij} do {
                        If (Library contains q_{ij}) {
                                If (current values for the q_{ij} in the Library < values for
                                this q_{ij}) then update the values for q_{ij} in the Library
                                with the newer one
                        }
                        else
                           Add q_{ij} to the Library
                }
        }
        Return Library;
}
```

Fig. 16 Pseudo code for the algorithm to create common query library

One important consequence of this methodology is that this framework can be applied to any other user models for IR or different types of target applications in the same logic. We can determine a set of attributes describing a typical user for that target application, and determine a set of attributes describing the target application and the effectiveness function to assess a user's current goal in using the target application. We can convert it to a decision theoretic framework if the domains for these attributes are finite. The elicitation process is domain or application-dependent.

We also contribute to both communities by reusing the collections, metrics from the IR community such as CACM and MEDLINE; and simulating the traditional procedure as well as creating new procedure for evaluating new features offered by this user model. By reusing the metrics, collections, and simulating the traditional procedure, we ensure comparability. By creating a new procedure, we ensure that we can further evaluate the newly added features from this technique.

Even though the framework for the hybrid user model has been created and substantively implemented and tested, additional work still needs to be done towards the goal of fully integrating both the user and system sides. In the current evaluation framework for the hybrid user model, we cannot evaluate the sub value function for threshold because it only takes the first 15 documents for relevancy assessment and the metric is average precision at three point fixed recalls. For evaluating the sub value function over threshold, a usability evaluation is needed so that the interactions between a user and the IR system are recorded. This can be done implicitly by using additional equipment, recording and processing a user's mouse clicks and gazes. Lastly, an experiment to study the behavior of our hybrid model and how it relates to the certain characteristics of a document collection is planned.

References

1. Allen, R.: User models: theory, method and practice. Int. J. Man Mach. Stud. **32**, 511–543 (1990)
2. Baeza-Yates, R., Calderón-Benavides, L., Gonzalez-Caro, C.: The intention behind Web queries. In: Proceedings of String Processing and Information Retrieval 2006, pp. 98–109. Glasgow, Scotland (2006)
3. Baeza-Yates, R., Raghavan, P.: Next generation Web search. In: Ceri, S., Brambilla, M. (eds.) Search Computing. Lecture Notes in Computer Science, vol. 5950, pp. 11–23. Springer, Berlin (2010)
4. Baeza-Yates, R., Ribiero-Neto, B.: Modern Information Retrieval. Addison-Wesley, New York (1999)
5. Balabanovic, M.: Exploring versus exploiting when learning user models for text recommendation. User Model. User-Adap. Inter. **8**(1–2), 71–102 (1998)
6. Balabanovic, M., Shoham, Y.: Content-based collaborative recommendation. Commun. ACM **40**(3), 66–72 (1997)
7. Belkin. N.J.: Interaction with text: information retrieval as information seeking behavior. Information retrieval. 10. von der Modelierung zur Anwerdung, pp. 55–66. Universitaetsverlag, Konstanz (1993)
8. Belkin, N.J., Oddy, R.N., Brooks, H.M.: Ask for information retrieval: part I: background and theory. J. Doc. **38**(2), 61–71 (1982)
9. Belkin, N.J. Windel, G.: Using monstrat for the analysis of information interaction. In: IRFIS 5, Fifth International Research Forum in Information Science, pp. 359–382. Heidelberg (1984)
10. Billsus, D., Pazzani, M.J.: User modeling for adaptive news access. User Model. User-Adap. Inter. **10**(2–3), 147–180 (2000)
11. Bodoff, D., Raban, D.: User models as revealed in web-based research services. J. Am. Soc. Inform. Sci. Technol. **63**(3), 584–599 (2012)
12. Borlund, P.: The concept of relevance in information retrieval. J. Am. Soc. Inform. Sci. Technol. **54**(10), 913–925 (2003)
13. Boughanem, M., Tmar, M.: Incremental adaptive filtering: profile learning and threshold calibration. In: Proceedings of SAC 2002, pp. 640–644. Madrid, Spain (2002)
14. Brajnik, G., Guida, G., Tasso, C.: User modeling in intelligent information retrieval. Inf. Process. Manage. **23**(4), 305–320 (1987)
15. Broder, A.: A taxonomy of Web search. SIGIR Forum **36**(2), 3–10 (2002)
16. Brown, S.M.: Decision theoretic approach for interface agent development. Ph.D. thesis, Air Force Institute of Technology (1998)
17. Campbell, I., van Rijsbergen, C.J.: Ostensive model of developing information needs. In: Proceedings of the Second International Conference on Conceptions of Library and Information Science: Integration in Perspective (CoLIS 2), pp. 251–268 (1996)
18. Cecchini, R.L., Lorenzetti, C.M., Maguitman, A.G., Brignole, N.B.: Using genetic algorithms to evolve a population of topical queries. Inf. Process. Manage. **44**(6), 1863–1878 (2008)
19. Chen, S.Y., Magoulas, G.D., Dimakopoulos, D.: A flexible interface design for Web directories to accommodate different cognitive styles. J. Am. Soc. Inform. Sci. Technol. **56**(1), 70–83 (2005)
20. Cooper, W., Maron, M.E.: Foundations of probabilistic and utility theoretic indexing. J. Assoc. Comput. Mach. **25**(1), 67–80 (1978)
21. Donghee, Y.: Hybrid query processing for personalized information retrieval on the Semantic Web. Knowl.-Based Syst. **27**, 211–218 (2012)
22. Drucker, H., Shahrary, B., Gibbon, C.: Support vector machines: relevance feedback and information retrieval. Inf. Process. Manage. **38**(3), 305–323 (2002)
23. Ducheneaut, N., Partidge, K., Huang, Q., Price, B., Roberts, M., Chi, E.H., Belotti, V., Begole, B.: Collaborative filtering is not enough? Experiments with a mixed-model

recommender for leisure activities. In: Proceeding of the Seventeenth International Conference, User Modeling, Adaptation, and Personalization, pp. 295–306. Trento, Italy (2009)

24. Duong H.T., Uddin, M.N., Lim D., Jo, G.: A collaborative ontology-based user profiles system. In: Computational Collective Intelligence. Semantic Web, Social Networks and Multiagent Systems, pp. 540–552 (2009)

25. Efthimis, E.N.: Query expansion. In: Williams, M. (ed.) Ann Rev Inf Sci Technol **31**, 121–187 (1996)

26. Frake, W.B., Baeza-Yates, R.: Information retrieval: data structures and algorithms, p. 07458. Prentice Hall PTR, Upper Saddle River (1992)

27. Ghorab M.R., Zhou D., O'Connor A., Wade, V.: Personalised information retrieval: survey and classification. User Modeling and User-Adapted Interaction. Online first (2012)

28. He, B., Ounis, I.: Inferring query performance using pre-retrieval predictors'. In: Information Systems, Special Issue for the String Processing and Information Retrieval: 11th International Conference, pp. 43–54 (2004)

29. Ide, E.: New experiment in relevance feedback. In: The Smart System Experiments in Automatic Documents Processing, pp. 337–354 (1971)

30. Ingwersen, P.: Information Retrieval Interaction. Taylor Graham, London (1992)

31. Jansen, B., Booth, D., Spink, A. Determining the user intent of Web search engine queries. In: Proceedings of the International World Wide Web Conference, pp. 1149–1150. Alberta, Canada (2007)

32. Jensen, F.V.: An Introduction to Bayesian Networks. University College London Press, London (1996)

33. Keeney, L.R., Raiffa, H.: Decision with Multiple Objectives: Preferences and Value Tradeoffs. Wiley, New York (1976)

34. Kim, J.: Describing and predicting information-seeking behavior on the Web. J. Am. Soc. Inform. Sci. Technol. **60**(4), 679–693 (2009)

35. Kofler, C., Lux, M.: Dynamic presentation adaptation based on user intent classification. In: Proceedings of the 17th ACM International Conference on Multimedia (MM '09), pp. 1117–1118. ACM, New York, USA (2009)

36. Kumaran, G., Allan, J.: Adapting information retrieval systems to user queries. Inf. Process. Manage. **44**(6), 1838–1862 (2008)

37. Lau, T., Horvitz, E.: Patterns of search: analyzing and modeling Web query refinement. In: Proceedings of the Seventh International Conference on User Modeling, pp. 119–128. Banff, Canada (1999)

38. Lee, U., Liu, Z., Cho, J.: Automatic identification of user goals in web search. In: Proceedings of the International World Wide Web Conference 2005, pp. 391–400. Chiba, Japan (2005)

39. Logan, B., Reece, S., Sparck, J.: Modeling information retrieval agents with belief revision. In: Proceedings of the Seventeenth Annual ACM/SIGIR Conference on Research and Development in Information Retrieval, pp. 91–100 (1994)

40. Lopér-Pujalte, C., Guerrero-Bote, V., Moya-Anegon, F.D.: Genetic algorithms in relevance feedback: a second test and new contributions. Inf. Process. Manage. **39**(5), 669–697 (2003)

41. Lynch, C.: The next generation of public access information retrieval systems for research libraries: lessons from ten years of the MELVYL system. Inf. Technol. Libr. **11**(4), 405–415 (1992)

42. Mat-Hassan, M., Levene, M.: Associating search and navigation behavior through log analysis. J. Am. Soc. Inform. Sci. Technol. **56**(9), 913–934 (2005)

43. Michard, M.: Graphical presentation of boolean expressions in a database query language: design notes and an ergonomic evaluation. Behav. Inf. Technol. **1**(3), 279–288 (1982)

44. Miller, G.A.: WordNet: a lexical database for English. Commun. ACM **38**(11), 39–41 (1995)

45. Nguyen, H.: Capturing user intent for information retrieval. Ph.D. Dissertation, University of Connecticut (2005)

46. Nguyen, H., Haddawy, P.: The decision-theoretic interactive video advisor. In: Proceedings of the Fifteenth Conference on Uncertainty in Artificial Intelligence- UAI 99, pp. 494–501. Stockholm, Sweden (1999)
47. Nguyen, H., Santos, E. Jr, Schuet, A., Smith, N.: Hybrid user model for information retrieval. In: Technical Report of Modeling Others from Observations workshop at Twenty-First National Conference on Artificial Intelligence (AAAI) conference, pp. 61–68. Boston (2006)
48. Nguyen, H., Santos, E.J., Zhao, Q., Lee, C.: Evaluation of effects on retrieval performance for an adaptive user model. In: Adaptive Hypermedia 2004 Workshop Proceedings—Part I, pp. 193–202., Eindhoven, The Netherlands (2004a)
49. Nguyen, H., Santos, E.J., Zhao, Q., Wang, H.: Capturing user intent for information retrieval. In: Proceedings of the Human Factors and Ergonomics Society 48th Annual Meeting, pp. 371–375. New Orleans, LA (2004b)
50. Pearl, J.: Probabilistic reasoning in intelligent systems: networks of plausible inference. Morgan Kaufmann, San Mateo (1988)
51. Rich, E.: User modeling via stereotypes. Cogn. Sci. 3, 329–354 (1979)
52. Rich, E.: Users are individuals: individualizing user models. Int. J. Man Mach. Stud. 18, 199–214 (1983)
53. Rochio, J.J.: Relevance feedback in information retrieval. In: The Smart system— experiments in automatic document processing, pp. 313–323 (1971)
54. Rose, D., Levinson, D.: Understanding user goals in Web search. In: Proceedings of the International World Wide Web Conference 2004, pp. 13–19. New York, USA (2004)
55. Ruthven, I., Lalmas, M.: A survey on the use of relevance feedback for information access systems. Knowl. Eng. Rev. 18(2), 95–145 (2003)
56. Ruthven, I., Lalmas, M., van Rijsbergen, K.: Incorporating user search behavior into relevance feedback. J. Am. Soc. Inform. Sci. Technol. 54(6), 529–549 (2003)
57. Salton, G., Buckley, C.: Improving retrieval performance by relevance feedback. J. Am. Soc. Inf. Sci. 41(4), 288–297 (1990)
58. Santos, E.J., Nguyen, H.: Modeling users for adaptive information retrieval by capturing user intent. In: Chevalier, M., Julien, C., Soulé, C. (eds.) Collaborative and Social Information Retrieval and Access: Techniques for Improved User Modeling, pp. 88–118. IGI Global (2009)
59. Santos, E.J., Nguyen, H., Brown, S.M.: Kavanah: an active user interface for information retrieval application. In: Proceedings of 2nd Asia-Pacific Conference on Intelligent Agent Technology, pp. 412–423, Japan (2001)
60. Santos, E.J., Nguyen, H., Zhao, Q., Pukinskis, E.: Empirical evaluation of adaptive user modeling in a medical information retrieval application. In: Proceedings of the Ninth User Modeling Conference, pp. 292–296, Johnstown (2003a)
61. Santos, E.J., Nguyen, H., Zhao, Q., Wang, H.: User modeling for intent prediction in information analysis. In: Proceedings of the 47th Annual Meeting for the Human Factors and Ergonomics Society (HFES-03), pp. 1034–1038, Denver (2003b)
62. Saracevic, T.: Relevance reconsidered. In: Ingwersen, P., Pors, P.O. (eds.) Proceedings of the Second International Conference on Conceptions of Library and Information Science: Integration in Perspective. Copenhagen: The Royal School of Librarianship, pp. 201–218 (1996)
63. Saracevic, T., Spink A., Wu, M.: Users and intermediaries in information retrieval: what are they talking about? In: Proceedings of the Sixth International Conference in User Modeling - UM 97, pp. 43–54 (1997)
64. Sleator, D.D., Temperley D.: Parsing English with a link grammar. In: Proceedings of the Third International Workshop on Parsing Technologies, pp. 277–292 (1993)
65. Spink, A., Cole, C.: New Directions in Cognitive Information Retrieval. The Information Retrieval Series. Springer (2005)
66. Spink, A., Losee, R.M.: Feedback in information retrieval. In: Williams, M. (ed.) Ann.Rev. Inf. Sci. Technol. 31, 33–78 (1996)

67. Spink, A., Greisdorf, H., Bateman, J.: From highly relevant to not relevant: examining different regions of relevance. Inf. Process. Manage. **34**(5), 599–621 (1998)
68. Steichen, B., Ashman, H., Wade, V.: A comparative survey of personalised information retrieval and adaptive hypermedia techniques. Inf. Process. Manage. **48**, 698–724 (2012)
69. Truran, M., Schmakeit, J., Ashman, H.: The effect of user intent on the stability of search engine results. J. Am. Soc. Inform. Sci. Technol. **62**(7), 1276–1287 (2011)
70. Vickery, A., Brooks, H.: Plexus: the expert system for referral. Inf. Process. Manage. **23**(2), 99–117 (1987)
71. Voorhees, M.E.: On test collections for adaptive information retrieval. Inf. Process. Manage. **44**(6), 1879–1885 (2008)
72. Xie, Y., Raghavan, V.V.: Language-modeling kernel based approach for information retrieval. J. Am. Soc. Inform. Sci. Technol. **58**(14), 2353–2365 (2007)
73. Zanker, M., Jessenitschnig, M.: Case-studies on exploiting explicit customer requirements in recommender systems. User Model. User-Adap. Inter. **19**(1–2), 133–166 (2009)
74. Zhang, Y.: Complex adaptive filtering user profile using graphical models. Inf. Process. Manage. **44**(6), 1886–1900 (2008)
75. Zhao, Q., Santos, E.J., Nguyen, H., Mohammed, M.: What makes a good summary? In: Argamon, S., Howard, N. (eds.) Computational Methods for Counterterrorism, pp. 33–50. Springer, New York, (2009)

Recommender Systems: Network Approaches

David Lamb, Martin Randles and Dhiya Al-Jumeily

Abstract The use of recommender systems is now common across the Web as users are guided to items of interest using prediction models based on known user data. In essence the user is shielded from information overload by being presented solely with data relevant to that user. Whilst this process is transparent, for the user, it transfers the burden of data analysis to an automated system that is required to produce meaningful results in real time from a huge amount of information. Traditionally structured data has been stored in relational databases to enable access and analysis. This chapter proposes the investigation of a new approach, to efficiently handle the extreme levels of information, based on a network of linked data. This aligns with more up-to-date methods, currently experiencing a surge of interest, loosely termed NoSQL databases. By forsaking an adherence to the relational model it is possible to efficiently store and reason over huge collections of unstructured data such as user data, document files, multimedia objects, communications, email and social networks. It is proposed to represent users and preferences as a complex network of vertices and edges. This allows the use of many graph-based measures and techniques by which relevant information and the underlying topology of the user structures can be quickly and accurately obtained. The centrality of a user, based on *betweenness* or *closeness*, is investigated using the Eigenvalues of the Laplacian spectrum of the generated graph. This provides a compact model of the data set and a measure of the relevance or importance of a particular vertex. Newly-developed techniques are assessed using Active Clustering and Acquaintance Nomination to identify the most influential participants in

D. Lamb (✉) · M. Randles · D. Al-Jumeily
School of Computing and Mathematical Sciences, Liverpool John Moores University,
Liverpool, UK
e-mail: D.J.Lamb@LJMU.ac.uk

M. Randles
e-mail: M.J.Randles@LJMU.ac.uk

D. Al-Jumeily
e-mail: D.AlJumeily@LJMU.ac.uk

G. A. Tsihrintzis et al. (eds.), *Multimedia Services in Intelligent Environments*,   65
Smart Innovation, Systems and Technologies 24, DOI: 10.1007/978-3-319-00372-6_4,
© Springer International Publishing Switzerland 2013

the network and so provide the best recommendations to general users based on a sample of their identified exemplars. Finally an implementation of the system is evaluated using real-world data.

1 Introduction

At present, the number and variety of information systems is continuing to grow and expand. This includes systems such as e-commerce applications, peer-to-peer networking, cloud storage, e-government, and mobile apps, which are expanding into many more areas of influence. These systems increasingly handle huge amounts of generated data, giving potential for many un-attenuated results in response to system user queries. Thus, users are not able to easily distinguish relevant contents from secondary concerns.

Recommender Systems (RS) are efficient tools designed to overcome this information overload problem by providing users with the most relevant content [26]. Recommendations are computed by predicting users' opinions (or ratings) for new content based on previous data regarding likes and dislikes and/or their similarity to existing users: collaborative recommendation. Rating predictions are usually based on a user profiling model that summarizes former users' behaviour. Current trends look towards aggregating far more user data into the user profiling process; including data from social media and mobile network services. Thus a user may have, or fit into, several profiles which may contain overlapping or redundant data. It is increasingly difficult to provide integrated recommendations within these vast interwoven data streams. It is therefore necessary to consider additional methods for data representation and handling, which more closely match the current interaction methods of users with recommender service systems.

As previously identified, there are a number of challenges in supplying recommender system services: most notably this firstly involves deciding on how to represent the users, their data and the related products/items, and how to do so in a scalable manner. Secondly it is necessary to extract individual recommendations from this data incorporating a number of approaches [14]. For this reason, it is proposed in this work to use a graph-based model to represent the user-product/item relationship and relevance. The model will consist of nodes, representing either customers or products, and links corresponding to transactions and similarities. In order to extract recommendations, queries can be constructed on known graph properties, applications and analysis techniques; making this approach an example of a NoSQL data store [17]. The move to NoSQL, initially founded upon the notion of Big Data, Big Storage and Cloud services [20]; intended to free developers from a reliance on relational databases and the SQL language. More recently, however, the term has come to stand for "Not Only SQL", emphasizing a best fit approach; providing alternative strategies to solve a problem that relational databases do not handle well. Recommender systems are just such a case, for at least the reasons outlined above.

In line with a NoSQL approach and using graph-based techniques a new recommender network-rewiring algorithm is developed. This, mediated by algebraic connectivity, provides a measured increase in the range and scope of observed recommendations to users. The dynamic nature of recommender systems means that network rewiring, as new selections, purchases, or recommendations are made, is a frequent occurrence in these systems.

For example a new purchase by a particular customer may trigger a series of new links (graph edges) to be created based on that customer's previous purchases. Likewise links may be deleted upon that customer poorly rating his purchase. In this work active clustering [9], a recently investigated technique, whereby like items are rewired together to provide a well-organized (easy to explore and/or query) structure is investigated.

The active clustering approach, originally used for load-balancing, is used in this work to test the application of the discussed techniques, whereby the recommendation clustering is mediated by algebraic connectivity; the clustering only proceeds if an increased algebraic connectivity of the recommender network is observed.

The work is organized as follows: Sect. 2 reviews the literature and shows the relevance of the graph-based techniques to recommender systems. Section 3 gives an overview of the tools and technologies to be utilized and defines the relevant measurement to be used in this work, to improve the recommender system, through the algebraic connectivity measure. The proposed method is detailed formally via experimentation in Sect. 4; results are assessed with a simulation showing the impact on recommender systems induced by an increasing algebraic connectivity of the recommender network. Section 5 discusses the results and other work in this area, while Sect. 6 concludes and looks at future work and further developments.

2 Recommender Systems Review

Recommender systems are distinguished from traditional data retrieval systems through the user/product interactive link and the processing of user data. User preferences are identified by monitoring a wide range of observable data, such as user features (age, sex, location, etc.), particular characteristics of items purchased or selected by the user and user behaviour.

Recommender systems first started to appear in the early 1990s [11]. The GroupLens system [24] monitored readers of Usenet news articles and recommended similar articles to similar users. To the present day recommender systems are now a common feature of e-commerce sites and online advertising to increase income for the business and alert their customers to products, of interest, that they may not have been unaware of. A widely used technique, particularly in the early recommender systems, was gathering explicit feedback from customers/users in the form of product ratings [3, 22]. This approach, generally termed Collaborative

Filtering, which is based on the user's previous evaluation of items or history of previous purchases, is hampered by insufficient user data in the early stages of product deployment (sparseness) or too much data for an established product (scalability) [8, 25]: When a new product is released it may not receive a recommendation level consummate with its relevance to users because of a low review rate. Conversely when a product becomes established, there may be too much data to form an appropriate recommendation. To alleviate this problem implicit feedback, garnered from the users' online behaviour (browsing history, system logs, etc.) may be used. This involves relating user, product and transaction data representations. This approach, usually termed Content Based Filtering, represents users as a set of associated items. This again has some problems as similarities are detected between items that share the same attributes or characteristics. Thus recommendations are only made for products very similar to those already known to the user [19].

Moving closer to a graph representation for user/item/transaction modelling, data mining techniques are often used to obtain user profiles, patterns of behaviour and establish rules to infer recommendations from profile data [4]. These involve multi-dimensional data, which may also include time; allowing targeted promotion based on the time of day/year/etc. [1]. It is, thus, possible to see how a graph based approach may be suitable for extracting recommendations using, for example, clustering, decision trees, k-nearest recommendation, link analysis or rule extraction linking one product with another.

Some work has been presented around these techniques, without necessarily explicitly relating graph techniques to the problems: self-organising maps can be used as a learning technique, for recommendations, through clustering [18]. Decision trees can be used to test attributes at non-leaf nodes with each branch the result of a test. The leaf nodes then form a class of recommendations [10]. The use of k-nearest recommendations technique involves establishing a user's preferences and identifying, through machine learning, users with similar preferences; these "recommendations" may then be used to establish additional recommendations [6]. Link analysis may be used to establish connections in social networks; algorithms are used in a similar way to PageRank to establish the major links in a network for enhanced recommendations [7]. For rule extraction, a data mining technique would establish a set of items A which would be linked to a set of items B, whereby $A => B$, meaning any time an item in set A is chosen, the items in set B are potentially relevant recommendations [16].

It can be seen from the foregoing discussion that multidimensional data and graphing techniques may be usefully applied to recommender systems. However, most existing systems rely on the analysis of the two dimensions of users and items and concentrate on specific types of graph. It is proposed in this work to investigate the specific mapping of data to a graph based representation, permitting unstructured data such as user data, document files, multimedia objects, communications, email and social network information to be included in recommender system outputs.

3 Background: Graphs and NoSQL

While the use of graph structures to represent recommender system data is not new [2, 21], this work seeks to investigate the use of modern graph analysis techniques for better recommendations. To this end a review of current NoSQL implementations will be completed and an active clustering algorithm, mediated by an increasing connectivity metric (range of recommendations) is proposed.

3.1 Current NoSQL Implementations

The emerging NoSQL movement has gained ground in the design of solutions to Big Data problems; particularly in the context of Cloud-based storage solutions and platforms. It aims to lessen the reliance on the often poor fit of relational databases for large-scale data persistence and query/access. The concepts and structures for these NoSQL alternatives are not particularly new; though the identification and adoption of alternative methods of data storage for common use is gaining popularity. Table 1 shows a small selection of current offerings.

These implementations represent a scalable alternative to traditional, relational databases—largely based on distributed hash tables and multidimensional data spaces [13]. It is clear, therefore, that such approaches have a graph representation as the underlying mechanism of storage and data retrieval.

This chapter seeks to engender increased quality recommendations by engineering robustness in the linked graph of customer/product or transaction/similarity representations, followed by beneficial adaptations instigated by random events at a local level. This proposal suggests engineering this robustness firstly through the optimisation of the connectivity measure (Algebraic Connectivity) at a meso-organisation level. Work at this level permits distribution of operation and computation, but still requires a level of global knowledge. Secondly, robustness is brought about by efficient random propagation; recommendation comparisons occur at the local level with increased quality recommendations emerging at the global level. These two strands are explained further in the next sections.

3.2 The Algebraic Connectivity Metric

At the meso-level in a recommender network system, underlying networks of components contribute to the formation of emergent recommendation structures.

For instance, in an idealised global music recommendation system, a single retailer's review site can be considered as a single recommender node on an Internet-scale music recommendation graph. Thus increasing the connectivity of an underlying recommender network ought to promote some beneficial effect; in

Table 1 Some typical NoSQL database implementations

	Company	Type	Storage	APIs	Reference
Big table	Google	Persistent multi-dimensional	Byte arrays	Phython, GQL, REST, etc.	http://labs.google.com/papers/bigtable-osdi06.pdf
BerkleyDB	Oracle	Key-value	B-Tree/hash table	C, C++, Java	http://www.oracle.com/database/dots/Berkley-DB-v-Relational.pdf
Cassandra	Apache	Hash table	Keyspaces	Java, Ruby, Perl, Phyton, C#	http://wiki.apache.org/cassandra/frontPage
Extreme scale	IBM	In memory grid	Java, persistent cache, map reduce	Java, REST	http://www.01.ibm.com/software/webservers/appserv/extremescale/library/index.html
MongoDB	10 gen	Open source, document	JSON	C, C++, Java, JavaScript, Perl, PHP, Phython, Ruby,C#, etc.	http://www.mongodb.org/display/DOCs/Documentation+index
SimpleDB	Amazon	Item/property/value	Rows/columns/multiple values	SOAP, REST	http://docs.amazonwebservices.com/AmazonsimpleDB/latest/DeveloperGuide/
Esent	Microsoft	Key-value	Columns/indices	Windows SDK	http://msdn.microsoft.com/en-us/library/5c485eff-4329-4dc1-aa45-fb66e654792.aspx

terms of identifying new recommendations and increasing the relevancy of such information.

In order to perform such reasoning it is necessary to portray the recommender networks in a manner that permits the extraction of suitable metrics. In this work particular matrices represent the connectivity of the recommendation networks: All matrices, of size $n \times n$, are assumed to be of the form:

$$\begin{pmatrix} a_{11} & \cdots & a_{1n} \\ \cdot & \cdot & \cdot \\ a_{n1} & \cdots & a_{nn} \end{pmatrix}$$

The *degree matrix* for the graph $G = <V, E>$ with $V = \{v_1, \ldots v_n\}$ and $E = \{(v_i, v_j) | v_i$ and v_j are linked$\}$ is given by:

$a_{ij} = \deg(v_i)$ if $i = j$ and 0 otherwise

The *adjacency matrix* is similarly given by:

$a_{ij} = 1$ if v_i is adjacent to v_j and 0 otherwise

The *admittance*, or *Laplacian matrix* is the symmetric, zero-row-sum matrix formed by subtracting the adjacency matrix from the degree matrix:

$a_{ij} = \deg(v_i)$ if $i = j$, -1 if v_i is adjacent to v_j and 0 otherwise

The spectrum of the graph G consists of the n Eigen values, $\lambda_0 \leq \lambda_1 \leq \ldots \leq \lambda_n$, of the Laplacian matrix. These are obtained by solving the characteristic equation $\det(L - \lambda I)$ det (L-λI) for λ where det is the matrix determinant, L is the Graph Laplacian and I is the identity matrix.

0 is always an Eigen value ($\lambda_0 = 0$ for all Laplacian matrices) and $\lambda_1 > 0$ is termed the algebraic connectivity. As the value of λ_1 increases, so does the connectivity of the graph; thus increasing the amount of clustering. This means that it is more difficult to split the graph into separate components; thus more recommendations become available for a given vertex.

It has been shown that the algebraic connectivity metric is a highly relevant measure in determining network connectivity and robustness [15]. In application to recommender systems, this work applies similar techniques in order to extract meaningful recommendations from less data; such as when part of the recommender network is not available.

The next section briefly discusses the approaches considered for efficient propagation of recommendations about the recommender graph.

3.3 Recommendation Comparison and Propagation

Network recommendations compare their recommending models as Markov Decision Problems. Recommendation models are assessed based on a reward value of the MDP, informed by the observer's global view (i.e. a user rating). If a component discovers that, based on this reward value, a recommendation component performs better; if capable, it can adapt itself to the improved recommender model.

In this way the benefits of self-organisation, autonomous recommendation optimisation and information updates are rapidly propagated across the system. It is assumed the components (individual or identified groups of customers or products) each control a Markov Decision Process (MDP) with the same underlying situation space, S. A component, i, has actions A_i and reward function R_i. The probability of a transition from situation s_1 to s_2, when action a is attempted, is denoted by: $p(s_2|s_1, a)$.

It is assumed that each component, i, implements a deterministic, stationary policy π_i, inducing a Markov Chain $p_i(s_2|s_1) = p_i(s_2|s_1,\pi_i(s_1))$. Also for all components, i and j, for each action $\pi_i(s)$ it is not necessarily the case that there exists an action $a \in A_j$ such that the distributions $p_j(.|s,a)$ and $p_i(.|s)$ are identical; thus a component can only emulate another component's model if it possesses the required functions. Thus for a component, i, gaining knowledge of another component, j's, transitions, though not necessarily the action that caused the transition, an augmented Bellman equation [23] for the value function, V, follows:

$$R_i + \gamma \max \left\{ \max_{a \in A_I} \left\{ \sum_{s_1 \in S} p_i(s_1|s, a) V(s_1) \right\}, \sum_{s_1 \in S} p_j(s_1|s) V(s_1) \right\}$$

where γ is a discount on future rewards. It should be noted that situations (action histories) are used instead of states. Adopting this approach allows the application of Situation Calculus logic, thus subjecting the domain to reasoning. Equally, the deliberation over quantified situations allows many domain instances to be considered at one time rather than addressing each individual domain instantiation. Recent work on exploiting the propositional nature of such problems [5] has brought about scalable techniques to provide solutions with domain state sizes of over 10^{40} [12].

4 The Effect of Algebraic Connectivity on Recommendations

In this example, a component consists of a network of linked products/customers. The component will have had specified a specific set of triggers for forced evolution; such as a low uptake of recommendations, new customers online or an explicit recommender network optimisation. This will enforce a graph rewiring, linking to another product or customer profile (edge realignment of the component network graph).

This rewiring is assessed and where it improves the component according to its associated Markov model, then it is retained. In this work the *recommendation availability* measure will be the algebraic connectivity of the component, whilst the effect of this on the value of the recommendation of the network containing the component will be measured.

In this section the strategy to promote relevant recommendations in a large network will be assessed. A two-stage strategy is effected to promote and increase *recommendation availability*, across and between products and consumers. The network will consist of products and customers, which each may comprise further networks of products and customers, indicative of a specific recommendation set.

Firstly, component networks will propagate recommender behaviour based on the reward value (utility) given by the Markov Decision Problem formulation. An example expression of the propagation strategy based on only local customer data is shown in pseudo-code for illustrative purposes.

```
Initialise customer, customerID, productList,
{recommendations}, utility
......
for each product in recommendation-list
{
   if (connection(customer, product)=="false") then
     recommendationList.remove(product)
   endif
}
if receive(ping) then
  newRecommendation(ping.CustomerID)=="true"
  for each product in productList
  {
     if (recommendation=ping.CustomerID)then
       newRecommendation(ping.CustomerID)=="false"
     endif
  }
  if (newRecommendation(ping.CustomerID)=="true") then
       newRecommendation==ping.customerID
       recommendationList.add(newRecommendation)
  endif
endif
for each product in recommendation-list
{
   if (utility <product.utility) then
     {recommendations}==product.{recommendations}
   endif
}........
```

Secondly, on the occurrence of a given evolution trigger (e.g. detection of low recommendation uptake), the network reconfigures, inducing the widely studied Small-World model, as a component searches amongst the separate components for a replacement recommendation set. At each potential reconfiguration the first non-zero Eigen value (Algebraic Connectivity) of the network graph is used as a measure of the suitability/*recommendation availability* of the new network topology. The reconfiguration only proceeds if a higher value is returned. Another pseudo-code snippet for this part of the recommender programme illustrates:

```
Initialise
Observer,customerNodeList,recommendationlist,
algebraicConn
.....
for each (aCustomerNode in customerList)
    {
    if (aCustomerNode.available!="true") then
       for each (bCustomerNode in customerList)
       {
          if (bCustomerNode.available.likeMe="true")  {BREAK}
       }
       connectionList.add((aCustomerNode, bCustomerNode))
       oldAlgebraicConn=algebraicConn
       calculate(algebraicConn)
       if (algebraicConn < oldAlgebraicConn) then
          recommendationList.remove(
                            (aCustomerNode, bCustomerNode))
       endif
    endif
    }.......
```

4.1 Application to Improve Recommendations

In [9] *Active Clustering* is considered as a self-aggregation algorithm to rewire networks. Application of this procedure is intended to group like-service instances together. This clustering is useful because many load-balancing and work-sharing algorithms only work well in cases where the nodes are aware of nodes similar to themselves and can easily delegate workload to them. Such an approach seems highly applicable to recommender systems where the search for like customers will produce new and relevant recommendations.

Active Clustering consists of iterative executions by each node in the network. In this case each customer/product object:

1. At a random time point the node becomes an "initiator" and selects a "matchmaker" node from its neighbours.
2. The "matchmaker" node searches for selects and causes a link to be formed between one of its neighbours that match the type of the "initiator" node and the "initiator" node.
3. The "matchmaker" removes the link between itself and its chosen neighbour.

This algorithm was studied extensively in [9], showing the organisation of a complex network towards a steady state.

In this work the "fast" algorithm does not allow the removal of a link between like nodes, whereas the "accurate" algorithm maintains the number of links and enforces that links between non-like nodes can only be added if another link between heterogeneous nodes is removed. Full details of the complete algorithm that switches between active fast and accurate as circumstances dictate may be found in [9].

It is noted that this algorithm provides improved results for load-balancing; there is, however, no consideration given to the effect on the network topology or the robustness of the rewired network. As such, these rewiring operations are largely considered in the context of a layer-separated "balancing" overlay.

In applying this algorithm to a recommender system network, the benefits of discovering new recommendations based on similarity are allied to increased recommendation availability. As such, the foregoing work, detailing the achievement of increased recommendations through algebraic connectivity mediated rewiring (clustering), is utilised. In this case *active clustering* is only enacted where the recommendations are increased thus eradicating the possibility that the strategy will compromise the recommender systems relevance.

5 Recommendations Experiment and Results

In order to compare the described algorithms an experiment was established using simulations. For simplicity the customer network is considered, where the product nodes form the link between like customers. A small sample-sized, scale-free network of 100 customers with 10 % heterogeneity (i.e. 10 different types of customers and product types) was used. The experiment was run using Active Clustering only, and then again with Algebraic Connectivity Mediated Active Clustering (ACMAC): For the ACMAC algorithm, the customer nodes iterate as normal but the system deliberation determines whether the rewiring (new recommendation) proceeds:

1. At a random time point a customer becomes an "initiator" and selects a "matchmaker" customer from its neighbours.
2. The "matchmaker" customer searches for selects and causes a link to be formed between one of its neighbours that match the type of the "initiator" customer and the customer.
3. The recommender system calculates the change in algebraic connectivity.
4. If the change is a positive value then the "matchmaker" removes the link between itself and its chosen neighbour, else the link between the initiator and the matchmaker's neighbour is removed.

Additionally for this experiment customer/recommendation failure (lack of complete data) is taken into account: in a first simulation, at a certain time point 25 % of the customers (and their recommendations) were randomly removed from the network. In the second simulation, 25 % of the recommendations were randomly removed from the network. The primary metric used here under these

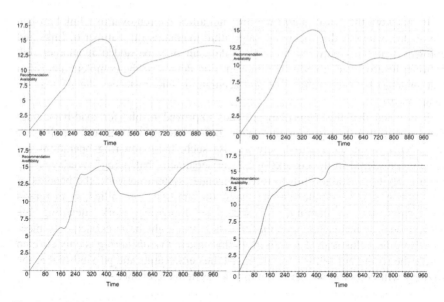

Fig. 1 Available recommendations for active clustering alone with 25 % node failure (*top-left*), algebraic connectivity mediated active clustering with 25 % node failure (*bottom-left*): active clustering alone with 25 % edge failure (*top-right*) and algebraic connectivity mediated active clustering with 25 % Edge failure (*bottom-right*)

experimental conditions is the availability of relevant recommendations; the number of completed recommendations per elapsed time unit. The simulations were repeated at least 20 times each.

The results of the simulations are shown in Fig. 1, revealing better long-term outcomes when algebraic connectivity is used with Active Clustering.

It is shown from the graphs in Fig. 1 that the initial recommendations on the network are slowed by the application of the system's algebraic connectivity algorithm. This is due to customers having to wait for system response before committing to the Active Clustering rewire.

When recommendation failure scenarios are introduced, however, it is clear that, where Active Clustering occurs only when it increases the algebraic connectivity of the recommendation network, then the availability of recommendations is improved: in the case of customer node failure, recommendation availability is affected but recovers to near pre-failure values, whilst in the case of product edge failure, the induced recommendations show almost no discernible effect. This is in contrast to the case where active clustering is not mediated by increasing algebraic connectivity. In this case recommendation availability falls and, although some recovery is made, recommendation levels are lower.

6 Conclusion

Recommender systems are now an integral part of shopping on the Web, and e-Commerce in general; to increase traffic or sales, providing recommendations where the customer is too busy to identify all products of interest and to help locate useful information. The representations and production of recommendations is usually performed on an ad-hoc domain-dependent basis with little regard for missing data or other applications.

In this work, a graph based approach has been suggested leveraging a proprietary NoSQL product together with novel techniques of analysing and organising graph based data. Customers, products, and transactions are modelled using graphs with nodes representing either customers or products and edges the transactions. The network is then optimized using a combination of active clustering to group like products and customers together and the algebraic connectivity measure to ensure that the connectivity (recommendation availability) is enhanced.

To provide an initial demonstration of the applicability of this method, an experiment was conducted whereby a significant proportion of the data for recommendations was not available: this graph-based approach showed that the availability of recommendations could still be maintained based on the proportion of the recommendation network remaining. Furthermore the quality of the recommendations is maintained by each customer node comparing, and updating where necessary, its own recommendation model using the utility within a Markov Decision Problem.

Much further work is required, not least in developing the notions required for utility. At present this utility is subjectively provided by customer ratings of recommendations. This could also be extended to consider the uptake rate of recommendations made, thus further qualifying recommendation methods with a reliability metric. Additionally, investigations into fully distributing the system observer will be of value.

References

1. Adomavicius, G., Tuzhilin, A.: Extending recommender systems: a multidimensional approach. In: Proceedings of the International Joint Conference on Artificial Intelligence (IJCAI-01), Workshop on Intelligent Techniques for Web Personalization (ITWP2001), Seattle, Washington (2001)
2. Aggarwal, C.C., Wolf, J.L., Wu, K.-L., Yu, P.S.: Horting hatches an egg: a new graph-theoretic approach to collaborative filtering. In: Proceedings of the Fifth ACM SIGKDD Conference on Knowledge Discovery and Data Mining (KDD'99), pp. 201–212, San Diego (1999)
3. Balabanovic, M., Shoham, Y.: Fab: content-based, collaborative recommendation. Commun. ACM **40**(3), 66–72 (1997)
4. Berry, M.J.A., Linoff, J.S.: Data mining techniques for marketing, sales and customer relationship management, 3rd edn, Wiley, Indiana (2004)

5. Boutilier C., Reiter, R. Price, R.: Symbolic dynamic programming for first-order MDPs. In: Proceedings of the Seventeenth International Joint Conference on Artificial Intelligence (IJCAI-01), Seattle, pp. 690–697 (2001)

6. Cacheda F., Carneiro, V., Fernández, D., Formoso, V.: Improving k-nearest neighbors algorithms: practical application of dataset analysis. In: Berendt, B., de Vries, A., Fan, W., Macdonald, C., Ounis, I., Ruthven, I. (eds.) Proceedings of the 20th ACM international conference on Information and knowledge management (CIKM '11), ACM, New York (2011)

7. Cai, D., He, X., Wen, J.-R., Ma, W.-Y.: Block-level link analysis. In: Proceedings of the 27th annual international ACM SIGIR conference on Research and development in information retrieval, pp. 440–447, ACM, (2004)

8. Claypool, M., Gokhale, A., Miranda, T., Murnikov, P., Netes, D., Sartin, M.: Combining content-based and collaborative filters in an online newspaper. In: Proceedings of the ACM SIGIR'99 workshop on recommender systems, (1999)

9. Di Nitto, E., Dubois, D.J., Mirandola, R., Saffre, F., Tateson, R.: Applying self-aggregation to load balancing: experimental results. In: Proceedings of the 3rd international Conference on Bio-inspired Models of Network, Information and Computing Systems (Bionetics 2008), Article 14, 25–28 November 2008

10. Golbandi N., Koren, Y., Lempel, R. (2011) Adaptive bootstrapping of recommender systems using decision trees. In: Proceedings of the Fourth ACM International Conference on Web Search and Data Mining (WSDM '11). ACM, New York, NY, USA

11. Goldberg, D., Nichols, D., Oki, B., Terry, D.: Using collaborative filtering to weave an information tapestry. Commun. ACM 35(12), 61–70 (1992)

12. Guestrin, C., Koller, D., Parr, R., Venkataraman, S.: Efficient solution algorithms for factored MDPs. J. Artif. Intell. Res. 19, 399–468 (2003)

13. Han J., Haihong, E., Le, G., Du, J.: Survey on NoSQL database. In: Proceedings of the 6th International Conference on Pervasive Computing and Applications (ICPCA), 2011, pp. 363–366, (2011)

14. Huang, Z., Chung, W., Chen, H.: A graph model for E-commerce recommender systems. J. Am. Soc. Inf. Sci. 55, 259–274 (2004)

15. Jamakovic, A., Van Mieghem, P.: On the robustness of complex networks by using the algebraic connectivity. In: Das et al A. (ed.) NETWORKING 2008 Ad Hoc and sensor networks, wireless networks, next generation internet, pp. 183–194, LNCS 4982, Springer (2008)

16. Jannach, D., Zanker, M., Felfernig, A., Friedrich, G.: Recommender systems: An introduction. Cambridge University Press, Cambridge (2011)

17. Leavitt, N.: Will NoSQL Databases Live Up to Their Promise? Computer 43(2), 12–14 (2010)

18. Lihua, W., Lu, L., Jing, L., Zongyong, L.: Modeling user multiple interests by an improved GCS approach. Expert Syst. Appl. 29, 757–767 (2005)

19. Lopez-Nores, M., Garca-Duque, J., Frenandez-Vilas, R.P., Bermejo-Munoz, J.: A flexible semantic inference methodology to reason about user preference in knowledge-based recommender systems. Knowl.-Based Syst. 21, 305–320 (2008)

20. Meng-Ju, H., Chao-Rui, C., Li-Yung, H., Jan-Jan, W., Pangfeng, L.: SQLMR : A scalable Database management system for cloud computing. In: Proceedings of the International Conference on Parallel Processing (ICPP), 2011, pp. 315–324, (2011)

21. Mirza, B.J.: Jumping connections: A graph-theoretic model for recommender systems. MSc. Dissertation, Faculty of the Virginia Polytechnic Institute and State University. http://scholar.lib.vt.edu/theses/available/etd-02282001-175040/unrestricted/etd.pdf. Accessed 13 July 2012 (2001)

22. Mooney, R., Roy, L.: Content-based book recommending using learning for text categorization. In: Proceedings of the Fifth ACM Conference on Digital Libraries, pp: 195–204, San Antonio, (2000)

23. Price R., Boutilier, C.: Imitation and reinforcement learning in agents with heterogeneous actions. In: Proceedings of the AISB'00 Symposium on Starting from Society—the Application of Social Analogies to Computational Systems, pp: 85–92, Birmingham, (2000)
24. Resnick, P., Iacovou, N., Suchak, M., Bergstrom, P., Riedl, J.: GroupLens: an open architecture for collaborative filtering of netnews. In: Proceedings of the 1994 ACM conference on Computer supported cooperative work (CSCW'94), pp. 175–186. ACM, New York, (1994)
25. Sarwar B., Karypis, G., Konstan, J.A., Riedl, J.: Analysis of recommendation algorithms for e-commerce. In: Proceedings of the ACM E-Commerce, pp. 158–167, (2000)
26. Schafer J.B., Konstan, J., Riedl, J.: Recommender Systems in E-Commerce. In: ACM Conference on Electronic Commerce, 1999, pp. 158–166

Resource List

This section contains a short list of related items useful for readers wishing to investigate further or learn more about the issues raised in this chapter:-

http://www.utm.edu/departments/math/graph/

(Graph Theory tutorials)

http://homepages.cae.wisc.edu/~gautamd/Active_Clustering/Home.html

(Basic overview of Active Clustering in Graph Theory)

http://martinfowler.com/articles/nosqlKeyPoints.html

(Overview of NoSQL)

http://nosql-database.org/

(Repository of NoSQL, key/value, and graph databases)

http://mappa.mundi.net/signals/memes/alkindi.shtml

("Alkindi" —Java Source code for commercial collaborative filtering recommender system, now unrestricted and no longer maintained)

http://mahout.apache.org/

(Mahout—Machine Learning Library, incorporating Recommender components. Part of the Apache project; maintained at the time of writing)

Toward the Next Generation of Recommender Systems: Applications and Research Challenges

Alexander Felfernig, Michael Jeran, Gerald Ninaus,
Florian Reinfrank and Stefan Reiterer

Abstract Recommender systems are assisting users in the process of identifying items that fulfill their wishes and needs. These systems are successfully applied in different e-commerce settings, for example, to the recommendation of news, movies, music, books, and digital cameras. The major goal of this book chapter is to discuss new and upcoming applications of recommendation technologies and to provide an outlook on major characteristics of future technological developments. Based on a literature analysis, we discuss new and upcoming applications in domains such as software engineering, data and knowledge engineering, configurable items, and persuasive technologies. Thereafter we sketch major properties of the next generation of recommendation technologies.

1 Introduction

Recommender systems support users in the identification of items that fulfill their wishes and needs. As a research discipline, *recommender systems* has been established in the early 1990s (see, e.g., [1]) and since then has grown enormously

A. Felfernig (✉) · M. Jeran · G. Ninaus · F. Reinfrank · S. Reiterer
Institute for Software Technology, Graz University of Technology,
Inffeldgasse 16b, A-8010 Graz, Austria
e-mail: alexander.felfernig@ist.tugraz.at

M. Jeran
e-mail: mjeran@ist.tugraz.at

G. Ninaus
e-mail: gerald.ninaus@ist.tugraz.at

F. Reinfrank
e-mail: florian.reinfrank@ist.tugraz.at

S. Reiterer
e-mail: stefan.reiterer@ist.tugraz.at

G. A. Tsihrintzis et al. (eds.), *Multimedia Services in Intelligent Environments*,   81
Smart Innovation, Systems and Technologies 24, DOI: 10.1007/978-3-319-00372-6_5,
© Springer International Publishing Switzerland 2013

in terms of algorithmic developments as well as in terms of deployed applications. A recommender system can be defined as a system that guides users in a personalized way to interesting or useful objects in a large space of possible objects or produces such objects as output [2, 3]. Practical experiences from the successful deployment of recommendation technologies in e-commerce contexts (e.g., www.amazon.com [4] and www.netflix.com [5]) contributed to the development of recommenders in new application domains. Especially such new and upcoming application domains are within the major focus of this chapter.

On an algorithmic level, there exist four basic recommendation approaches. First, *content-based filtering* [6] is an information filtering approach where features of items a user liked in the past are exploited for the determination of new recommendations. Content-based filtering recommendation is applied, for example, by www.amazon.com for the recommendation of items which are similar to those that have already been purchased by the user. If a user has already purchased a book related to the *Linux* operating system, new (and similar) books will be recommended to her/him in the future.

Second, *collaborative filtering* is applied to the recommendation of items such as music and movies [4, 7, 8]. It is based on the concept of analyzing the preferences of users with a similar item rating behavior. As such, it is a basic implementation of word-of-mouth promotion with the idea that a purchase decision is influenced by the opinions of friends and relatives. For example, if user *Joe* has purchased movies similar to the ones that have been purchased by user *Mary* then www.amazon.com would recommend items to *Joe* which have been purchased by *Mary* but not by *Joe* up to now. The major differences compared to content-based filtering are (a) no need of additional item information (in terms of categories or keywords describing the items) and (b) the need of information about the rating behavior of other users.

Third, high-involvement items such as cars, apartments, and financial services are typically recommended on the basis of *knowledge-based recommendation technologies* [2, 9]. These technologies are based on the idea of exploiting explicitly defined recommendation knowledge (defined in terms of deep recommendation knowledge, e.g., as rules and constraints) for the determination of new recommendations. Rating-based recommendation approaches such as collaborative filtering and content-based filtering are not applicable in this context since items are not purchased very frequently. A consequence of this is that no up-to-date rating data of items is available. Since knowledge-based recommendation approaches rely on explicit knowledge representations the so-called knowledge acquisition bottleneck becomes a major challenge when developing and maintaining such systems [9].

Finally, *group recommendation* [10] is applied in situations where there is a need of recommendations dedicated to a group of users, for example, movies to be watched by a group of friends or software requirements to be implemented by a development team. The major difference compared to the afore mentioned recommendation approaches lies in the criteria used for determining recommendations: while in the case of content-based filtering, collaborative filtering, and knowledge-

Table 1 Overview of journals and conferences (including workshops) used as the basis for the literature analysis (papers on recommender system applications 2005–2012)

Journals and conferences
International Journal of Electronic Commerce (IJEC)
Journal of User Modeling and User-Adapted Interaction (UMUAI)
AI Magazine
IEEE Intelligent Systems
Communications of the ACM
Expert Systems with Applications
ACM Recommender Systems (RecSys)
User Modeling, Adaptation, and Personalization (UMAP)
ACM Symposium on Applied Computing (ACM SAC)
ACM International Conference on Intelligent User Interfaces (IUI)

based recommendation the major goal is to identify recommendations that perfectly fit the preferences of the current user, group recommendation approaches have to find ways to satisfy the preferences of a user group.

The major focus of this book chapter is to give an overview of new and upcoming applications of recommendation technologies and to provide insights into major requirements regarding the development of the next generation of recommender systems. Our work is based on an analysis of research published in workshops, conferences, and journals summarized in Table 1. In this context we do not attempt to provide an in-depth analysis of state-of-the-art recommendation algorithms [11, 12] and traditional applications of recommendation technologies [13] but focus on the aspect of new and upcoming applications [3, 14]. An overview of these applications is given in Table 2.

The remainder of this chapter is organized as follows. In Sects. 2–6 we provide an overview of new types of applications of recommendation technologies and give working examples. In Sect. 7 we sketch the upcoming generation of recommender systems as *Personal Assistants* which significantly improve the overall quality of recommendations in terms of better taking into account preferences of users—in this context we discuss major issues for future research.

2 Recommender Systems in Software Engineering

Recommender systems can support stakeholders in software projects by efficiently tackling the information overload immanent in software projects [15]. They can provide stakeholder support throughout the whole software development process—examples are the *recommendation of methodological knowledge* [16, 17], the *recommendation of requirements* [18, 19], and the *recommendation of code* [20, 21].

Table 2 Overview of identified recommender applications not in the mainstream of e-commerce applications (Knowledge-based Recommendation: KBR, Collaborative Filtering: CF, Content-based Recommendation: CBR, Machine Learning:ML, Group Recommendation: GR, Probability-based Recommendation: PR, Data Mining: DM)

Domain	Recommended items	Recommendation approach	
Software engineering	Effort estimation methods	KBR	Peischl et al. [17]
	Problem solving approaches	KBR	Burke and Ramezani [16]
	API call completions	CF	McCarey et al. [21]
	Example code fragments	CBR	Holmes et al. [22]
	Contextualized artifacts	KBR	Kersten and Murphy [23]
	Software defects	ML	Misirli et al. [24]
	Requirements prioritizations	GR	Felfernig et al. [18]
	Software requirements	CF	Mobasher and Cleland-Huang [19]
Data and knowledge engineering	Related constraints	KBR	Felfernig et al. [26, 27]
	Explanations (hitting sets)	KBR	Felfernig et al. [28]
	Constraint rewritings	KBR	Felfernig et al. [29]
	Database queries	CF	Chatzopoulou et al. [30]
		CBF	Fakhraee and Fotouhi[31]
Knowledge-based configuration	Relevant features	CF	Felfernig and Burke [32]
	Requirements repairs	CBF	Felfernig et al. [33]
Persuasive technologies	Game task complexities	CF	Berkovsky et al. [34]
	Development practices	KBR	Pribik and Felfernig [35]
Smart homes	Equipment configurations	KBR	Leitner et al. [36]
	Smart home control actions	CF	LeMay et al. [37]
People	Criminals	PR	Tayebi et al. [38]
	Reviewers	CF	Kapoor et al. [39]
	Physicians	CF	Hoens et al. [40]
Points of interest	Tourism services	CF	Huang et al. [41]
	Passenger hotspots	PR	Yuan et al. [42]
Help services	Schools	CBR	Wilson et al. [43]
	Financial services	KBR	Fano and Kurth [44]
	Lifestyles	KBR	Hammer et al. [45]
	Receipes	CBR	Pinxteren et al. [46]
	Health information	CBR	Wiesner and Pfeifer [47]
Innovation management	Innovation teams	KBR	Brocco and Groh [48]
	Business plans	KBR	Jannach and Joergensen [49]
	Ideas	DM	Thorleuchter et al. [50]

Recommendation of Methodological Knowledge. Appropriate method selection is crucial for successful software projects, for example, the waterfall process is only applicable for risk-free types of projects but not applicable for high risk projects. Method recommendation is already applied in the context of different types of development activities such as the domain-dependent recommendation of algorithmic problem solving approaches [16] and the recommendation of appropriate effort estimation methods [17]. These approaches are based on the knowledge-based recommendation paradigm [2, 9] since recommendations do not rely on the taste of users but on well-defined rules defining the criteria for method selection.

Recommendation of Code. Due to frequently changing development teams, the intelligent support of discovering, locating, and understanding code is crucial for efficient software development [21]. Code recommendation can be applied, for example, in the context of (collaborative) call completion (which API methods from a large set of options are relevant in a certain context?) [21], recommending relevant example code fragments (which sequence of methods is needed in the current context?) [22], tailoring the displayed software artifacts to the current development context [23], and classification-based defect prediction [24]. For an in-depth discussion of different types of code recommendation approaches and strategies we refer the reader to [15, 25].

Recommendation of Requirements. Requirements engineering is one of the most critical phases of a software development process and poorly implemented requirements engineering is one of the major reasons for project failure [51]. Core requirements engineering activities are elicitation and definition, quality assurance, negotiation, and release planning [52]. All of these activities can be supported by recommendation technologies, for example, the (collaborative) recommendation of requirements to stakeholders working on similar requirements [19] and the group-based recommendation of requirements prioritizations [18].

Example: Content-based Recommendation of Similar Requirements. In the following, we will exemplify the application of *content-based filtering* [6] in the context of *requirements engineering*. A recommender can support stakeholders, for example, by recommending requirements that have been defined in already completed software projects (requirements reuse) or have been defined by other stakeholders of the same project (redundancy and dependency detection). Table 3 provides an overview of requirements defined in a software project. Each requirement req_i is characterized by a *category*, the number of estimated *person days* to implement the requirement, and a textual *description*.

Table 3 Example of a content-based filtering recommendation problem: recommendation of similar requirements based on *category* and/or *keyword* information

Requirement	Category	Person days	Description
req_1	Database	170	Store component configuration in DB
req_2	User interface	65	User interface with online help available
req_3	Database	280	Separate tier for DB independence
req_4	User interface	40	User interface with corporate identity

Table 4 Example of a group recommendation problem: recommendation of requirements prioritizations to a group of stakeholders (used heuristics = majority voting)

Requirement	stakeholder$_1$	stakeholder$_2$	stakeholder$_3$	stakeholder$_4$	Recommendation
req$_1$	1	1	1	2	1
req$_2$	5	4	5	5	5
req$_3$	3	3	2	3	3
req$_4$	5	5	4	5	5

If we assume that the stakeholder currently interacting with the requirements engineering environment has already investigated the requirement req_1 (assigned to the category *database*), a content-based recommender system would recommend the requirement req_3 (if this one has not been investigated by the stakeholder up to now). If no explicitly defined categories are available, the textual description of each requirement can be exploited for the extraction of keywords which serve for the characterization of the requirement. Extracted keywords can then be used for the determination of similar requirements [53]. A basic metric for determining the similarity between two requirements is given in Formula 1. For example, $sim(req_1, req_3) = 0.17$, if we assume $keywords(req_1)$= {*store, component, configuration, DB*} and $keywords(req_3)$ = {*tier, DB, independence*}.

$$sim(req_1, req_2) = \frac{|keywords(req_1) \cap keywords(req_2)|}{|keywords(req_1) \cup keywords(req_2)|} \tag{1}$$

Example: Group-based Recommendation of Requirements Prioritizations. Group recommenders include heuristics [10] that can help to find solution alternatives which will be accepted by the members of a group (with a high probability). Requirements prioritization is the task of deciding which of a given set of requirements should be implemented within the scope of the next software release—as such, this task has a clear need of group decision support: a stakeholder group has to decide which requirements should be taken into account. Different heuristics for coming up with a group recommendation are possible, for example, the *majority heuristic* proposes a recommendation that takes the majority of requirements-individual votes as group recommendation (see Table 4). In contrast, the *least misery* heuristic recommends the minimum of the requirements-individual ratings to avoid misery for individual group members. For a detailed discussion of different possible group recommendation heuristics we refer the reader to [10].

3 Recommender Systems in Data and Knowledge Engineering

Similar to conventional software development, knowledge-based systems development suffers from continuously changing organizational environments and personal. In this context, recommender systems can help database and knowledge engineers to better cope with the size and complexity of knowledge structures [30, 33]. For example, recommender systems can support an improved understanding of the knowledge base by actively suggesting those parts of the knowledge base which are candidates for increased testing and debugging [26, 33]. Furthermore, recommender systems can propose repair and refactoring actions for faulty and ill-formed knowledge bases [27, 29]. In the context of information search, recommender systems can improve the accessibility of databases (knowledge bases) by recommending queries [30].

Knowledge Base Understanding. Understanding the basic elements and the organizational structure of a knowledge base is a major precondition for efficient knowledge base development and maintenance. In this context, recommender systems can be applied for supporting knowledge engineers, for example, by (collaboratively) recommending constraints to be investigated when analyzing a knowledge base (recommendation of navigation paths through a knowledge base) or by recommending constraints which are related to each other, i.e., are referring to common variables [26].

Knowledge Base Testing and Debugging. Knowledge bases are frequently adapted and extended due to the fact that changes in the application domain have to be integrated into the corresponding knowledge base. Integrating such changes into a knowledge base in a consistent fashion is time-critical [9]. Recommender systems can be applied for improving the efficiency of knowledge base testing and debugging by recommending minimal sets of changes to knowledge bases which allow to restore consistency [28, 33]. Popular approaches to the identification of such *hitting sets* are model-based diagnosis introduced by Reiter [54] and different variants thereof [28].

Knowledge Base Refactoring. Understandability and maintainability of knowledge bases are important quality characteristics which have to be taken into account within the scope of knowledge base development and maintenance processes [55]. Low quality knowledge structures can lead to enormous maintenance costs and thus have to be avoided by all available means. In this context, recommender systems can be applied for recommending relevant refactorings, i.e., semantics-preserving structural changes of the knowledge base. Such refactorings can be represented by simple constraint rewritings [29] or by simplifications in terms of recommended removals of redundant constraints, i.e., constraints which do not have an impact on the semantics of the knowledge base [27].

Recommender Systems in Databases. Chatzopoulou et al. [30] introduce a recommender application that supports users when interactively exploring relational databases (*complex SQL queries*). This application is based on collaborative filtering where information about the navigation behavior of the current user is exploited for the recommendation of relevant queries. These queries are generated on the basis of the similarity of the current user's navigation behavior and the navigation behavior of nearest neighbors (other users with a similar navigation behavior who already searched the database). Fakhraee et al. [31] focus on recommending database queries which are derived from those database attributes that contain the keywords part of the initial user query. In contrast to Chatzopoulou

Table 5 Example of a collaborative recommendation problem. The entry ij with value 1 (0) denotes that fact that knowledge engineer$_i$ has (not) inspected constraint$_j$

	constraint$_1$	constraint$_2$	constraint$_3$	constraint$_4$
knowledge engineer$_1$	1	0	1	?
knowledge engineer$_2$	1	0	1	1
knowledge engineer$_3$	1	1	0	1

et al. [30], Fakhraee et al. [31] do not support the recommendation of complex SQL queries but focus on basic lists of keywords (*keyword-based database queries*).

Example: Collaborative Recommendation of Relevant Constraints. When knowledge engineers try to *understand a given set of constraints* part of a knowledge base, recommender systems can provide support in terms of showing related constraints, for example, constraints that have been analyzed by knowledge engineers in a similar navigation context. In Table 5, the constraints {constraint$_{1..4}$} have already partially been investigated by the knowledge engineers {knowledge engineer$_{1..3}$}. For example, knowledge engineer$_1$ has already investigated the constraints constraint$_1$ and constraint$_3$. Collaborative filtering (CF) can exploit the ratings (the rating = 1 if a knowledge engineer has already investigated a constraint and it is 0 if the constraint has not been investigated up to now) for identifying additional constraints relevant for the knowledge engineer.

User-based CF [7] identifies the k-nearest neighbors (knowledge engineers with a similar knowledge navigation behavior) and determines a prediction for the rating of a constraint the knowledge engineer had not to deal with up to now. This prediction can be determined, for example, on the basis of the majority of the k-nearest neighbors. In the example of Table 5 knowledge engineer$_2$ is the nearest neighbor (if we set k = 1) since knowledge engineer$_2$ has analyzed (or changed) all the constraints investigated by knowledge engineer$_1$. At the same time, knowledge engineer$_1$ did not analyze (change) the constraint constraint$_4$. In our example, CF would recommend the constraint$_4$ to knowledge engineer$_1$.

4 Recommender Systems for Configurable Items

Configuration can be defined as a basic form of design activity where a target product is composed from a set of predefined components in such a way that it is consistent with a set of predefined constraints [56]. Similar to knowledge-based recommender systems [2, 9], configuration systems support users in specifying their requirements and give feedback in terms of solutions and corresponding explanations [57]. The major difference between configuration and recommender systems is the way in which these systems represent the product (configuration) knowledge: configurators are exploiting a configuration knowledge base whereas (knowledge-based) recommender systems are relying on a set of enumerated solution alternatives (items). Configuration knowledge bases are especially useful in scenarios with a large number of solution alternatives which would make an explicit representation extremely inefficient [57].

In many cases, the amount and complexity of options presented by a configurator outstrips the capability of a user to identify an appropriate solution. Recommendation technologies can be applied in various ways to improve the usability of configuration systems, for example, filtering out product features which are relevant for the user [57, 58], proposing concrete feature values and thus helping the user in situations where knowledge about certain product properties is not available [57], and by determining plausible repairs for inconsistent user requirements [33].

Selecting Relevant Features. In many cases users are not interested in specifying all the features offered by a configurator interface, for example, some users may be interested in the GPS functionality of digital cameras whereas this functionality is completely uninteresting for other users. In this context, recommender systems can help to determine features of relevance for the user, i.e., features the user is interested to specify. Such a feature recommendation can be implemented, for example, on the basis of collaborative filtering [32].

Determining Relevant Feature Values. Users try to avoid to specify features they are not interested in or about which they do not have the needed technical knowledge [57]. In such a context, recommender systems can automatically recommend feature values and thus reduce the burden of interaction for the user. Feature value recommendation can be implemented, for example, on the basis of collaborative filtering [7, 57]. Note that feature value recommendation can trigger biasing effects and—as a consequence—could also be exploited for manipulating users to select services which are not necessarily needed [59].

Determining Plausible Repairs for Inconsistent Requirements. In situations where no solution can be found for a given set of customer requirements, configuration systems determine a set of repair alternatives which guarantee the recovery of consistency. Typically many different repair actions are determined by the configurator which makes it nearly infeasible for the user to find one which exactly fits his/her wishes and needs. In such a situation, knowledge-based and collaborative recommendation technologies can be exploited for personalizing the user requirements repair search process [33].

Example: Collaborative Recommendation of Features. Table 6 represents an interaction log which indicates in which session which features have been selected by the user (in which order)—in $session_1$, $feature_1$ was selected first, then $feature_3$ and $feature_2$, and finally $feature_4$. One approach to determine relevant features for (the current) user in $session_5$ is to apply collaborative filtering. Assuming that the user in $session_5$ has specified the $features_{1,2}$, the most similar session would be $session_2$ and $feature_3$ would be recommended to the user since it had been selected by the nearest neighbor (user in $session_2$). For a discussion of further feature selection approaches we refer the reader to [57].

Table 6 Example of collaboratively recommending relevant features. A table entry x denotes the order in which a user specified values for the given features

	feature$_1$	feature$_2$	feature$_3$	feature$_4$
session$_1$	1	3	2	4
session$_2$	1	2	3	4
session$_3$	1	1	4	3
session$_4$	1	3	2	4
session$_5$	1	2	?	?

5 Recommender Systems for Persuasive Technologies

Persuasive technologies [60] aim to trigger changes in a user's attitudes and behavior on the basis of the concepts of human computer interaction. The impact of persuasive technologies can be significantly increased by additionally integrating recommendation technologies into the design of persuasive systems. Such an approach moves persuasive technologies forward from a one-size-fits all approach to a personalized environment where user-specific circumstances are taken into account when generating persuasive messages [61]. Examples of the application of recommendation technologies in the context of persuasive systems are the enforcement of physical activity while playing computer games [34] and encouraging software developers to improve the quality of their software components [35].

Persuasive Games. Games focusing on the motivation of physical activities include additional reward mechanisms to encourage players to perform real physical activities. Berkovsky et al. [34] show the successful application of collaborative filtering recommendation technologies [7] for estimating the personal difficulty of playing. This recommendation (estimation) is exploited to adapt the difficulty level of the current game session since the perceived degree of difficulty is correlated with the preparedness of a user to perform physical activities.

Persuasive Software Development Environments. Software development teams are often under the gun of developing software components under high time pressure which often has a direct impact on the corresponding software quality. However, in the long term software quality is strongly correlated with the degree of understandability and maintainability. In this context, software quality improvements can be achieved by recommendation technologies, for example, knowledge-based recommenders can be applied to inform programmers about critical code segments and also recommend actions to be performed in order to increase the software quality. Pribik et al. [35] introduce such an environment which has been implemented as an Eclipse plugin (www.eclipse.org).

Example: Collaborative Estimation of Personal Game Level Difficulties. Table 7 depicts a simple example of the application of collaborative filtering to the determination of personal difficulty levels. Let us assume that the current user in session$_5$ has already completed the tasks 1..3; by determining the nearest neighbor of session$_5$ we can infer a probable duration of task$_4$: session$_1$ is the nearest

Table 7 Collaborative filtering based determination of personal game difficulty levels. A table entry x denotes the time a user needed to complete a certain game task

	task$_1$	task$_2$	task$_3$	task$_4$
session$_1$	4	6	5	7
session$_2$	1	2	2	2
session$_3$	4	5	5	6
session$_4$	1	1	1	2
session$_5$	4	6	4	?

neighbor of session$_5$ and the user of session$_1$ needed seven time units to complete task$_4$. This knowledge about the probable time efforts needed to complete a task can be exploited to automatically adapt the complexity level of the game (with the goal to increase the level of physical activity [34]).

6 Further Applications

Up to now we discussed a couple of new and innovative application domains for recommendation technologies. Admittedly, this enumeration is incomplete since it does not reflect wide-spread e-commerce applications—for a corresponding overview the interested reader is referred to [4, 13]. In this section we discuss further new and upcoming application domains for recommendation technologies that have been identified within the scope of our literature analysis.

Recommender Systems for Smart Homes. Smart homes are exploiting information technologies to improve the quality of life inside the home. Leitner et al. [36] show the application of knowledge-based recommendation technologies in the context of ambient assisted living (AAL) scenarios where on the one hand recommenders are applied to the design a smart home, on the other hand to control the smart home equipment. During the design phase of a smart home, the user is guided through a preference construction process with the final outcome of a recommendation of the needed technical equipment for the envisioned smart home. For already deployed smart home installations recommendation technologies support the people living in the smart home by recommending certain activities such as activating the air conditioning or informing other relatives about dangerous situations (e.g., for some relevant time period the status of the elderly people living in the house is unclear). A further application of recommendation technologies within the context of AAL is reported by LeMay et al. [37] who introduce an application of collaborative recommendation technologies for supporting the control of complex smart home installations.

People Recommender. Social networks such as *facebook.com* or *linkedin.com* are increasingly popular communication platforms. These platforms also include different types of recommender applications that support users in retrieving interesting music, travel destinations, and connecting to other people. Recommendations determined by these platforms are often exploiting the information contained in the underlying social network [62]. A new and upcoming application which exploits the information of social networks is the identification of crime suspects. In this context, social networks are representing the relationships between criminals. The *CrimeWalker* system introduced by Tayebi et al. [38] is based on a random-walk based method which recommends (provides) a list of the top-k potential suspects of a crime. Similar to *CrimeWalker*, *TechLens* is a recommender system which focuses on the recommendation of persons who could act—for example—as reviewers within the scope of a scientific conference organization. In contrast to *CrimeWalker*, the current version of *TechLens* does not exploit information from social networks—it is based on a collaborative filtering recommendation approach. Yuan et al. [42] present their approach to the recommendation of passenger hotspots, i.e., the recommender system is responsible for improving the prediction quality of hotspots and with this decreases the idle time of taxi drivers. Finally, Hoens et al. [40] present their approach to the recommendation of physicians—the major goal behind this application is to provide mechanisms which improve the quality of physician selection which otherwise is based on simple heuristics such as the opinion of friends or the information found on websites.

RFID based Recommenders. Personalized services become ubiquitous, for example, in tourism many destinations such as museums and art galleries are providing personalized access to help customers to better cope with the large amount of available services. An alternative to force users to explicitly declare their preferences before receiving a recommendation of potentially interesting services is to observe a user's navigation behavior and—on the basis of this information—to infer plausible recommendations. Such a recommendation approach is in the need of location and object information in order to be able to store the user navigation behavior. A collaborative filtering based handheld guide for art museums is introduced in [41] where Radio-frequency Identification (RFID) serves as a basis for preference data acquisition. RFID is a non-contact tracking system which exploits radio-frequency electromagnetic fields to transfer tag data for the purposes of object identification and object tracking [63].

Help Agents. Howto's are an important mechanism to provide guidance for users who are non-experts in a certain problem domain—such a support can be implemented, for example, on the basis of recommendation technologies [64]. One example for such a type of *help agent* is *SmartChoice* which is a (content-based) recommender system that supports representatives of low-income families in the choice of a public school for their children. Such a recommendation support is crucial since in many cases (a) parents do not dispose of detailed knowledge about the advantages and disadvantages of the different school types and (b) a false decision can have a very negative impact on the future life of the children. Another example of a help agent is *Personal Choice Point* [44] which is a financial service recommendation environment focusing on the visualization of the impact of different financial decisions on a user's life. Hammer et al. [45] present *MED-StyleR* which is a lifestyle recommender dedicated to the support of diabetes patients with the goal of improving care provision, enhancing the quality of a patient's life, and also to lower costs of public health institutions and patients. Another lifestyle related recommender application is presented by Pinxteren et al. [46] which focuses on determining health-supporting recipes that also fit the lifestyle of the user. In the same line, Wiesner and Pfeifer [47] introduce a content-based recommender application dedicated to the identification of relevant health information for a specific user.

Recommender Systems in Open Innovation. Integrating consumer knowledge into a company's innovation processes (also denoted as Open Innovation [65]) is in many cases crucial for efficient new product and service development. Innovation process quality has a huge impact on the ability of a company to achieve sustainable growth. Innovations are very often triggered by consumers who are becoming active contributors in the process of developing new products. Platforms such as *sourceforge.net* or *ideastorm.com* confirm this trend of progressive customer integration. These platforms exploit community knowledge and preferences to come up with new requirements, ideas, and products. In this context, the size and complexity of the generated knowledge (informal descriptions of requirements and ideas as well as knowledge bases describing new products on a formal level) outstrips the capability of community members to retain a clear view. Recommender systems can provide help in terms of finding other users with similar interests and ideas (*team recommendation* [48]) and to (semi-) automatically filter out the most promising ideas (*idea mining* [49, 50]).

7 Issues for Future Research

Recommendation technologies have been successfully applied for almost two decades primarily with the goal of increasing the revenue of different types of online services. In most of the existing systems the primary focus is to help to achieve business goals, rarely the viewpoint of the customer is taken into account in the first place [66]. For example, www.amazon.com recommenders inform users about new books of interest for them; a more customer-centered recommendation approach would also take into account (if available) information about books that have been bought by friends or relatives and thus are potentially available for the customer [66]. As a consequence, we are in the need of new recommendation technologies that allow more customer-centered recommendation approaches. In this context, the following research challenges have to tackled.

Focusing on the User Perspective. There are many other scenarios quite similar to the above mentioned www.amazon.com one where the recommender system is clearly focused on increasing business revenues. For example, consumer packaged goods (CPG) are already offered on the basis of recommender systems [67], however, these systems are domain-specific, i.e., do not take into account information regarding goods and services offered by the grocer nearby. Digital camera recommenders recommend the newest technology but in most cases do not take into account the current portfolio of the user, for example, if a user has a complete lens assortment of camera provider X it does not make sense to recommend a new camera of provider Y in the first place. An approach which is in the line of the idea of a stronger focus on the quality of user support is the RADAR personal assistant introduced by Faulring et al. [68] that supports multi-task coordination of personal emails.

Sharing Recommendation Knowledge. Besides commercial interests, one of the major reasons for the low level of customer orientation of todays recommender solutions is the lack of the needed recommendation knowledge. In order to recommend books already read by friends the recommender would need the information of the social network of the customer. The global availability of CPG goods information seems to be theoretically possible but is definitely in the need of a corresponding cloud and mobile computing infrastructure. More customer-centered recommender systems will follow the paradigm of personal assistants which does not focus on specific recommendation services but rather provides an integrated and multi-domain recommendation service [66, 69]. Following the idea of *ambient intelligence* [70], such systems will be based on global object information [63, 71] and support users in different application contexts in a cross-domain fashion.

Context Awareness. New recommendation technologies will intensively exploit the infrastructure of mobile services to determine and take into account the context of the current user [11]. Information such as the users shorttime and longterm preferences, geographical position, movement data, calendar information, information from social networks can be exploited for detecting the current context of the person and exploit this information for coming up with intelligent recommendations [72].

Unobtrusive Preference Identification. Knowledge about user preferences is a key preliminary for determining recommendations of relevance for the user. A major issue in this context is the development of new technologies which allow the elicitation of preferences in an unobtrusive fashion [73–75]. The three major modalities to support such a type of preference elicitation are the detection of facial expressions, the interpretation of recorded speech, and the analysis of physiological signals. An example of the derivation of user preferences from the analysis of eye tracking patterns is presented by Xu et al. [76] who exploit eye tracking technologies by interpreting attention times to improve the quality of a content-based filtering recommender. An approach to preference elicitation from physiological signals is presented by Janssen et al. [77] who exploit the information about skin temperature for measuring valence which is applicable to mood measurement.

Psychological Aspects of Recommender Systems. Building efficient recommendation algorithms and the corresponding user interfaces requires a deep understanding of human decision processes. This goal can be achieved by analyzing existing psychological theories of human decision making and their impact on the construction of recommender systems. Cosley et al. [78] already showed that the style of item rating presentation has a direct impact on a users' rating behavior. Adomavicius et al. [79] showed the existence of anchoring effects in different collaborative filtering scenarios. With their work, Adomavicius et al. [79]

confirm the results presented by Cosley et al. [78] but they show in more detail in which way rating drifts can have an impact on the rating behavior of a user. As already mentioned, recommendation technologies improve the quality of persuasive interfaces [34, 35]. Future recommenders should exploit the information provided by the mentioned preference elicitation methods [77].

8 Conclusions

With this chapter we provide an overview of new and upcoming applications of recommendation technologies. This overview does not claim to be complete but is the result of an analysis of work published in recommender systems related workshops, conferences, and journals. Beside providing insights into new and upcoming applications of recommendation technologies we also provide a discussion of issues for future research with the goal of advancing the state of the art in recommender systems which is characterized by a more user-focused and personal assistance based recommendation paradigm.

References

1. Goldberg, D., Nichols, D., Oki, B., Terry, D.: Using collaborative filtering to weave an information tapestry. Commun. ACM 35(12), 61–70 (1992)
2. Burke, R.: Knowledge-based recommender systems. Encycl. Libr. Inf. Syst. 69(32), 180–200 (2000)
3. Konstan, J., Riedl, J.: Recommender systems: from algorithms to user experience. User Model. User-Adap. Inter. 22(1), 101–123 (2012)
4. Linden, G., Smith, B., York, J.: Amazon.com recommendations—item-to-item collaborative filtering. IEEE Internet Comput. 7(1), 76–80 (2003)
5. Tuzhilin, A., Koren, Y.: In: Proceedings of the 2nd KDD Workshop on Large-Scale Recommender Systems and the Netflix Price Competition, pp. 1–34 (2008)
6. Pazzani, M., Billsus, D.: Learning and revising user profiles: the identification of interesting web sites. Mach. Learn. 27, 313–331 (1997)
7. Konstan, J., Miller, B., Maltz, D., Herlocker, J., Gordon, L., Riedl, J.: GroupLens: applying collaborative filtering to Usenet news. Commun. ACM 40(3), 77–87 (1997)
8. Koren, Y., Bell, R., Volinsky, C.: Matrix factorization techniques for recommender systems. IEEE Comput. 42(8), 30–37 (2009)
9. Felfernig, A., Friedrich, G., Jannach, D., Zanker, M.: An integrated environment for the development of knowledge-based recommender applications. Int. J. Electron. Commer. 11(2), 11–34 (2006)
10. Masthoff, J.: Group recommender systems: combining individual models. Recommender Systems Handbook, pp. 677–702. Springer, New York (2011)
11. Adomavicius, G., Tuzhilin, A.: Toward the next generation of recommander systems: a survey of the state-of-the-art and possible extensions. IEEE Trans. Knowl. Data Eng. 17(6), 734–749 (2005)
12. Jannach, D., Zanker, M., Felfernig, A., Friedrich, G.: Recommender Systems—An Introduction. Cambridge University Press, Cambridge (2010)

13. Schafer, J., Konstan, J., Riedl, J.: E-commerce recommendation applications. J. Data Min. Knowl. Discov. **5**(1–2), 115–153 (2011)
14. Ducheneaut, N., Patridge, K., Huang, Q., Price, B., Roberts, M.: Collaborative filtering is not enough? experiments with a mixed-model recommender for leisure activities. In: 17th International Conference User Modeling, Adaptation, and Personalization (UMAP 2009), pp. 295–306. Trento (2009)
15. Robillard, M., Walker, R., Zimmermann, T.: Recommendation systems for software engineering. IEEE Softw. **27**(4), 80–86 (2010)
16. Burke, R., Ramezani, M.: Matching recommendation technologies and domains. Recommender Systems Handbook, pp. 367–386. Springer, New York (2010)
17. Peischl, B., Zanker, M., Nica, M., Schmid, W.: Constraint-based recommendation for software project effort estimation. J. Emerg. Technol. Web Intell. **2**(4), 282–290 (2010)
18. Felfernig, A., Zehentner, C., Ninaus, G., Grabner, H., Maalej, W., Pagano, D., Weninger, L., Reinfrank, F.: Group decision support for requirements negotiation. Springer Lect. Notes Comput. Sci. **7138**, 1–12 (2011)
19. Mobasher, B., Cleland-Huang, J.: Recommender systems in requirements engineering. AI Mag. **32**(3), 81–89 (2011)
20. Cubranic, D., Murphy, G., Singer, J., Booth, K.: Hipikat: a project memory for software development. IEEE Trans. Softw. Eng. **31**(6), 446–465 (2005)
21. McCarey, F., Cinneide, M., Kushmerick, N.: Rascal—a recommender agent for agile reuse. Artif. Intell. Rev. **24**(3–4), 253–273 (2005)
22. Holmes, R., Walker, R., Murphy, G.: Approximate structural context matching: an approach to recommend relevant examples. IEEE Trans. Softw. Eng. **32**(12), 952–970 (2006)
23. Kersten, M., Murphy, G.: Using task context to improve programmer productivity. In: 14th ACM SIGSOFT International Symposium on Foundations of, Software Engineering, pp. 1–11, New York (2010)
24. Misirli, A., Bener, A., Kale, R.: AI-based software defect predictors: applications and benefits in a case study. AI Mag. **32**(2), 57–68 (2011)
25. Happel, H., Maalej, W.: Potentials and challenges of recommendation systems for software engineering. In: International Workshop on Recommendation Systems for Software Engineering, pp. 11–15. Atlanta (2008)
26. Felfernig, A., Reinfrank, F., Ninaus, G.: Resolving anomalies in feature models. In: 20th International Symposium on Methodologies for Intelligent Systems, pp. 1–10. Macau (2012)
27. Felfernig, A., Zehentner, C., Blazek, P.: CoreDiag: Eliminating redundancy in constraint sets. In: 22nd International Workshop on Principles of Diagnosis, Munich (2011)
28. Felfernig, A., Schubert, M., Zehentner, C.: An efficient diagnosis algorithm for inconsistent constraint sets. Artif. Intell. Eng. Des. Anal. Manuf. **25**(2), 175–184 (2011)
29. Felfernig, A., Mandl, M., Pum, A., Schubert, M.: Empirical knowledge engineering: cognitive aspects in the development of constraint-based recommenders. In: 23rd International Conference on Industrial, Engineering and Other Applications of Applied Intelligent Systems (IEA/AIE 2010), pp. 631–640. Cordoba (2010)
30. Chatzopoulou, G., Eirinaki, M., Poyzotis, N.: Query recommendations for interactive database exploration. In: 21st International Conference on Scientific and Statistical Database, Management, pp. 3–18 (2009)
31. Fakhraee, S., Fotouhi, F.: TupleRecommender: a recommender system for relational databases. In: 22nd International Workshop on Database and Expert Systems Applications (DEXA), pp. 549–553. Toulouse (2011)
32. Felfernig, A., Burke, R.: Constraint-based recommender systems: technologies and research issues. In: 10th ACM Int Conference on Electronic Commerce (ICEC'08), pp. 17–26. Innsbruck (2008)
33. Felfernig, A., Friedrich, G., Schubert, M., Mandl, M., Mairitsch, M., Teppan, E.: Plausible repairs for inconsistent requirements. In: IJCAI'09, pp. 791–796. Pasadena (2009)
34. Berkovsky, S., Freyne, J., Coombe, M., Bhandari, D.: Recommender algorithms in activity motivating games. In: ACM Conference on Recommender Systems (RecSys'09), pp. 175–182 (2010)

35. Pribik, I., Felfernig, A.: Towards persuasive technology for software development environments: an empirical study. In: Persuasive Technology Conference (Persuasive 2012), pp. 227–238. Linkoping (2012)
36. Leitner, G., Fercher, A., Felfernig, A., Hitz, M.: Reducing the entry threshold of AAL systems: preliminary results from casa vecchia. In: 13th International Conference on Computers Helping People with Special Needs, pp. 709–715. Linz (2012)
37. LeMay, M., Haas, J., Gunter, C.: Collaborative recommender systems for building automation. In: Hawaii International Conference on System Sciences, Waikoloa, Hawaii, 2009, pp. 1–10. Hawaii (2009)
38. Tayebi, M., Jamali, M., Ester, M., Glaesser, U., Frank, R.: Crimewalker: a recommender model for suspect investigation. In: ACM Conference on Recommender Systems (RecSys'11), pp. 173–180. Chicago (2011)
39. Kapoor, N., Chen, J., Butler, J., Fouty, G., Stemper, J., Riedl, J., Konstan, J.: Techlens: a researcher's desktop. In: 1st Conference on Recommender Systems, pp. 183–184. Minneapolis (2007)
40. Hoens, T., Blanton, M., Chawla, N.: Reliable medical recommendation systems with patient privacy. In: 1st ACM International Health Informatics Symposium (IHI 2010), pp. 173–182. Arlington (2010)
41. Huang, Y., Chang, Y., Sandnes, F.: Experiences with RFID-based interactive learning in museums. Int. J. Auton. Adapt. Commun. Syst. **3**(1), 59–74 (2010)
42. Yuan, N., Zheng, Y., Zhang, L., Xie, X.: T-Finder: A Recommender System for Finding Passengers and Vacant Taxis. IEEE Transactions on Knowledge and Data Engineering (TKDE) pp. 1–14 (2012)
43. Wilson, D., Leland, S., Godwin, K., Baxter, A., Levy, A., Smart, J., Najjar, N., Andaparambil, J.: Smartchoice: an online recommender system to support low-income families in public school choice. AI Mag. **30**(2), 46–58 (2009)
44. Fano, A., Kurth, S.: Personal choice point: helping users visualize what it means to buy a BMW. In: 8th International Conference on Intelligent User Interfaces (IUI 2003), pp. 46–52. Miami (2003)
45. Hammer, S., Kim, J., Andr, E.: MED-StyleR: METABO diabetes-lifestyle recommender. In: 4th ACM Conference on Recommender Systems, pp. 285–288. Barcelona (2010)
46. Pinxteren, Y., Gelijnse, G., Kamsteeg, P.: Deriving a recipe similarity measure for recommending healthful meals. In: 16th International Conference on Intelligent User Interfaces, pp. 105–114. Palo Alto (2011)
47. Wiesner, M., Pfeifer, D.: Adapting recommender systems to the requirements of personal health record systems. In: 1st ACM International Health Informatics Symposium (IHI 2010), pp. 410–414. Arlington (2010)
48. Brocco, M., Groh, G.: Team recommendation in open innovation networks. In: ACM Conference on Recommender Systems (RecSys'09), pp. 365–368. New York (2009)
49. Jannach, D., Bundgaard-Joergensen, U.: SAT: a web-based interactive advisor for investor-ready business plans. In: International Conference on e-Business (ICE-B 2007), pp. 99–106 (2007)
50. Thorleuchter, D., VanDenPoel, D., Prinzie, A.: Mining ideas from textual information. Expert Syst. Appl. **37**(10), 7182–7188 (2010)
51. Hofmann, H., Lehner, F.: Requirements engineering as a success factor in software projects. IEEE Softw. **18**(4), 58–66 (2001)
52. Sommerville, I.: Software Engineering. Pearson, Upper Saddle River (2007)
53. Mooney, R., L.R.: Content-based book recommending using learning for text categorization. User Model. User-Adap. Inter. **14**(1), 37–85 (2004)
54. Reiter, R.: A theory of diagnosis from first principles. AI J. **23**(1), 57–95 (1987)
55. Barker, V., O'Connor, D., Bachant, J., Soloway, E.: Expert systems for configuration at digital: XCON and beyond. Commun. ACM **32**(3), 298–318 (1989)
56. Sabin, D., Weigel, R.: Product configuration frameworks—a survey. IEEE Intell. Syst. **14**(4), 42–49 (1998)

57. Falkner, A., Felfernig, A., Haag, A.: Recommendation technologies for configurable products. AI Mag. **32**(3), 99–108 (2011)
58. Garcia-Molina, H., Koutrika, G., Parameswaran, A.: Information seeking: convergence of search, recommendations, and advertising. Commun. ACM **54**(11), 121–130 (2011)
59. Mandl, M., Felfernig, A., Tiihonen, J., Isak, K.: Status Quo Bias in configuration systems. In: 24th International Conference on Industrial Engineering and Other Applications of Applied Intelligent Systems (IEA/AIE 2011), pp. 105–114. Syracuse (2011)
60. Fogg, B.: Persuasive Technology—Using Computers to Change What We Think and Do. Morgan Kaufmann Publishers, Burlington (2003)
61. Berkovsky, S., Freyne, J., Oinas-Kukkonen, H.: Influencing individually: fusing personalization and persuasion. ACM Trans. Interact. Intell. Syst. **2**(2), 1–8 (2012)
62. Golbeck, J.: Computing with Social Trust. Springer, Heidelberg (2009)
63. Thiesse, F., Michahelles, F.: Building the internet of things using RFID. IEEE Internet Comput. **13**(3), 48–55 (2009)
64. Terveen, L., Hill, W.: Beyond recommender systems: helping people help each other. In: HCI in the New Millennium, pp. 487–509. Addison-Wesley (2001)
65. Chesbrough, H.: Open Innovation: The New Imperative for Creating and Profiting from Technology. Harvard Business School Press, Cambridge (2003)
66. Martin, F., Donaldson, J., Ashenfelter, A., Torrens, M., Hangartner, R.: The big promise of recommender systems. AI Mag. **32**(3), 19–27 (2011)
67. Dias, M., Locher, D., Li, M., El-Deredy, W., Lisboa, P.: The value of personalized recommender systems to e-business. In: 2nd ACM Conference on Recommender Systems (RecSys'08), pp. 291–294. Lausanne (2008)
68. Faulring, A., Mohnkern, K., Steinfeld, A., Myers, B.: The design and evaluation of user interfaces for the RADAR learning personal assistant. AI Mag. **30**(4), 74–84 (2009)
69. Chung, R., Sundaram, D., Srinivasan, A.: Integrated personal recommender systems. In: 9th ACM International Conference on Electronic Commerce, pp. 65–74. Minneapolis (2007)
70. Ramos, C., Augusto, J., Shapiro, D.: Ambient intelligence—the next step for artificial intelligence. IEEE Intell. Syst. **23**(2), 15–18 (2008)
71. Ramiez-Gonzales, G., Munoz-Merino, P., Delgado, K.: A collaborative recommender system based on space-time similarities. IEEE Pervasive Comput. **9**(3), 81–87 (2010)
72. Ballatore, A., McArdle, G., Kelly, C., Bertolotto, M.: RecoMap: an interactive and adaptive map-based recommender. In: 25th ACM Symposium on Applied Computing (ACM SAC 2010), pp. 887–891. Sierre(2010)
73. Foster, M., Oberlander, J.: User preferences can drive facial expressions: evaluating an embodied conversational agent in a recommender dialog system. User Model. User-Adap. Inter. **20**(4), 341–381 (2010)
74. Lee, T., Park, Y., Park, Y.: A time-based approach to effective recommender systems using implicit feedback. Expert Syst. Appl. **34**(4), 3055–3062 (2008)
75. Winoto, P., Tang, T.: The role of user mood in movie recommendations. Expert Syst. Appl. **37**(8), 6086–6092 (2010)
76. Xu, S., Jiang, H., Lau, F.: Personalized online document, image and video recommendation via commodity eye-tracking. In: ACM Conference on Recommender Systems (RecSys'08), pp. 83–90. New York (2008)
77. Janssen, J., Broek, E., Westerink, J.: Tune in to your emotions: a robust personalized affective music player. User Model. User-Adap. Inter. **22**(3), 255–279 (2011)
78. Cosley, D., Lam, S., Albert, I., Konstan, J., Riedl, J.: Is seeing believing how recommender system interfaces affect users opinions. In: CHI03, pp. 585–592 (2003)
79. Adomavicius, G., Bockstedt, J., Curley, S., Zhang, J.: Recommender systems, consumer preferences, and anchoring effects. In: RecSys 2011 Workshop on Human Decision Making in Recommender Systems, pp. 35–42 (2011)

Content-Based Recommendation for Stacked-Graph Navigation

Alejandro Toledo, Kingkarn Sookhanaphibarn,
Ruck Thawonmas and Frank Rinaldo

Abstract In this chapter we present a system which combines interactive visual analysis and recommender systems to support insight generation for the user. Collaborative filtering is a common technique for making recommendations; however, most collaborative-filtering systems require explicit user ratings and a large amount of existing data on each user to make accurate recommendations. In addition, these systems often rely on predicting what users will like based on their similarity to other users. Our approach is based on a content-based recommender algorithm, where promising stacked-graph views can be revealed to the user for further analysis. By exploiting both the current user navigational data and view properties, the system allows the user to see unseen-views suggested by our system. After testing with more than 30 users, we analyze the results and show that accurate user profiles can be generated based on user behavior and view property data.

A. Toledo (✉)
Laboratorio Nacional de Informática Avanzada, LANIA, Veracruz, Mexico
e-mail: atoledo@lania.mx

K. Sookhanaphibarn
Bangkok University, Bangkok, Thailand
e-mail: kingkarn@bu.ac.th

R. Thawonmas · F. Rinaldo
Ritsumeikan University, Kyoto, Japan
e-mail: ruck@ci.ritsumei.ac.jp

F. Rinaldo
e-mail: rinaldo@is.ritsumei.ac.jp

G. A. Tsihrintzis et al. (eds.), *Multimedia Services in Intelligent Environments*,
Smart Innovation, Systems and Technologies 24, DOI: 10.1007/978-3-319-00372-6_6,
© Springer International Publishing Switzerland 2013

1 Introduction

The rapid growth of online information has led to the development of recommenders systems, software tools providing suggestions for items to be of interest to a user. Online retailers, such as Amazon.com succesfully operate with different types of recommender systems. These tools have proved to be a valuable means for coping with the information overload problem. In addition to providing suggestions, recommender systems also apply personalization, considering that different users have different preferences and different information needs [1, 2].

In order to produce suggestions, recommender systems must collect personal preference information, e.g., the user's history of purchases, click-stream data, demographic information, etc. Traditionally, these expressions are called *ratings*. Two different types of ratings are distinguished, *explicit* and *implicit*. Explicit ratings are user explicit preferences for any particular item, usually by indicating their extent of appreciation on 5-point likert scales, which are mapped into numerical values. Negative values commonly indicate low appreciation, while positive values express some degree of appreciation.

Explicit ratings impose additional effort on users. Therefore, they often tend to avoid explicitly stating their preferences. Alternatively, collecting preference information from implicit observations of user behaviors is much less obstrusive [3–5]. Moreover, they are easier to collect. Typical examples for implicit ratings are purchased data, reading time, mouse activity, etc. The primary advantage to using implicit ratings is that such ratings remove the cost to the user of providing feedback [6].

In this chapter, we present a system which combines information visualization and recommender systems to support information seeking tasks in interactive visual analysis. In our system, a stacked graph visualization employs a content-based recommendation algorithm which infers user preferences from view dwell times and view properties. After a usability study with 32 subjects, we show that effective guidance in visual search space can be produced by analyzing the time users spent on views, the properties of those views, and the properties of a collection of views obtained in advance.

2 Related Work

Our work is similar to a number of projects whose effort has led to a collection of visual analytic tools that help users explore, analyze, and synthesize data [7–11]. However, much of the work therein has focused primarily on helping users visualize and interact with data sets. In our work, we aim to support visual exploration of existing data together with an approach for automatically guiding users in the task of exploring time-series data sets. To this end, a recommender approach selects candidate views of a stacked graph for proposing to the users, aiming to match their analytical needs.

The notion of whether a visualization system can suggest useful views is an intriguing idea that has not been explored much. Koop et al. [12] proposed Vis-Complete, a system that aids users in the process of creating visualizations by using previously created visualization pipelines. The system learns common paths used in existing pipelines and predicts a set of likely module sequences that can be presented to the user as suggestions during the design process. This method is similar to the predictions made by unix command lines and can be considered as using the collaborative filtering approach of recommender systems. Our work uses the content-based approach. More recently, Crnovrsanin et al. [13] developed an interactive network visualization system that suggests directions for further navigation based on past user interaction and calculated node importance.

Gotz et al. [14] proposed interactive tools to manage both the existing information and the synthesis of new analytic knowledge for sense-making in visualization systems. This work so far has not paid much attention on how to consolidate the users discoveries. Yang et al. [15] present an analysis-guided exploration system that supports the user by automatically identifying important data nuggets based on the interests of the users.

The notion of Implicit/Explicit feedback has been used to support recommendation systems. Traditional relevant feedback methods require that users explicitly or implicitly give feedbacks about their usage behaviors. Claypool et al. [3] provided a categorization of different implicit interest categories, such as, dwell times on a page, mouse clicks, and scrolling. Studies on dwell times, in particular, have gained a special interest in the research community [4]. Morita and Shinoda [5] studied dwell times in the form of reading times to explore how behaviors exhibited by users while reading articles could be used as implicit feedback.

Parson et al. [16] evaluated dwell times as an indicator of preference for attributes of items. In his work, Parson identifies a variety of factors that potentially could affect the dwell time, particularly in an uncontrolled setting. We believe that dwell times, as an indicator of preference, can be reliably extracted during interactive visual analysis. This is because viewers are arguably more focused than, for instance, buyers purchasing goods online. The most notable difference between Parson's method and our method is the application domain. Parson conducted experiments on a e-commerce setting; our method is based, and evaluated, on a visual analytics setting.

Related work on stacked graphs can be found in web sites such as Sense.us [17], ManyEyes [8], and NameVoyager [18], which allow many users to create, share, and discuss visualizations. One key feature of these systems is that they leverage the knowledge of a large group of people to effectively understand disparate data. Havre et al. [19] used the stacked graph to depict thematic variations over time within a large collection of documents. There is also work aiming at improving the aesthetics of the stacked graph [20, 21]. In our work, we adapt a recommender engine to the stacked graph visualization.

3 Stacked Graphs

In this work, we use and extend the stacked graph visualization proposed in [17]. The stacked graph is a practical method for visualizing changes in a set of time series, where the sum of their item values is as important as the individual values. Our stacked graph implementation has been designed to visualize different time series sets. Here, we use a collection of 500 job occupations reported in the United States labor force from 1850 to 2000.

The method used to visualize the data is straightforward, i.e., given a set of occupation time series, a stacked graph visualization is produced, as shown in Fig. 1. The horizontal axis corresponds to years and the vertical axis to occurrence ratios, in percentage against all occupations, for the occupations currently in view. Each stripe represents an occupation name, and the width of the stripe is proportional to the ratio of that occupation in a given year. The stripes are colored blue and pink for male and female genders, respectively. The brightness of each stripe varies according to the number of occurrences; the occupation with the largest population, for the whole period, is darkest and stands out the most.

As shown in Fig. 1, when the system starts, the user sees a set of stripes representing all occupation names with the population range between 100 and 52,609,716. The former corresponds to the occupation with the lowest number of population (Professor of Statistics), and the latter to the occupation with the highest number (Manager).

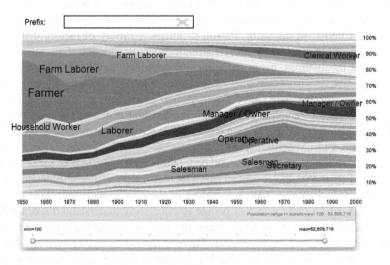

Fig. 1 Stacked graph showing a collection of 500 occupations

3.1 Views and View Properties

In the stacked graph, visual exploration is usually performed by data filtering. This allow users to produced the so called *views*, collections of stripes resulting from filtering effects. In our system, filtering is achieved using two interaction controls. With the first one, filtering by a prefix, the user may type in letters, forming a prefix; our system will then display a view showing occupation names beginning with that prefix.

With the second interaction control, filtering by a population range, the user can change the data currently in use from the default values. The system provides a slider allowing the change using any population range between 100 and 52,609,716. The idea behind this interaction control is that we can restrict the view to certain data of interest, according to their population range, resulting in concise views of the data. Figure 2 shows a stacked graph filtered by both a prefix and a population range.

All possible prefixes and population ranges serve as view attributes. Prefix values are regarded as textual attributes, whereas population ranges as numerical attributes. In the following, A denotes the set of all possible attributes in a view. $\alpha \subseteq A$ denotes the textual-attribute subset, and $\beta \subseteq A$ denotes the numerical-attribute subset, where $\alpha \cap \beta \equiv \emptyset$ and $\alpha \cup \beta \equiv A$.

A view v is defined by a tuple of textual and numerical properties as $v = (\alpha(v), \beta(v)) \in V$, where V denotes the view data set, of time series of interest. $\alpha(v)$ is the tuple of textual attribute values of view v, and $\beta(v)$ the tuple of numerical attribute values of view v.

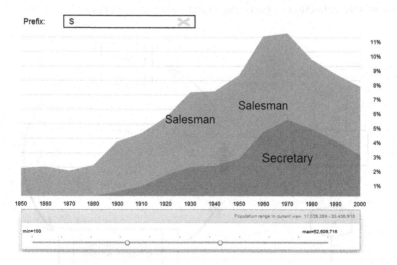

Fig. 2 Stacked graph filtered by a prefix and a population range

4 Content-Based Recommendation

To produce recommendations, we propose a solution based on the DESIRE algorithm proposed by Parson [16]. Here, we use three components as shown in Fig. 3. The first component is used to produce a *view data set*, which is a collection of views extracted from a stacked graph. The second component produces *user profiles* from user navigation data. User profiles consist of collections of unseen views, as well as user-preferences for such views. The last component produces *content-based recommendations*, i.e., collections of potentially useful unseen views, which are recommended for further analysis. Eventually, the unseen-views may become seen-views, so the process is repeated accordingly while the system is being used. In the following sections, we provide a detailed description of each of these three components.

4.1 View Data Set

As part of an off-line computation, we obtained what we call the *view data set*, which is the collection of all unique views a user can produce from the system using the two aforementioned interaction controls. Considering the huge number of prefixes and population ranges, and their possible combinations, it soon becomes clear that the total number of unique views would produce an extremely large view data set, thus impairing the system performance. One approach of producing a smaller view data set is considering prefixes of shorter length and a representative collection of population ranges. Using this approach, we obtained a

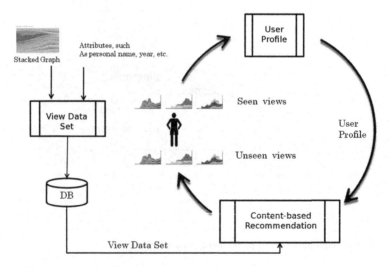

Fig. 3 Content-based recommendation

Table 1 Scheme structure of the view data set

View id	Textual attributes (prefix)					Numerical attributes (population range)			
	a	b	c	...	z	min	max	z(min)	z(max)
1	1	0	0	...	0	100	1000	−1.18	−0.77
2	0	0	1	...	0	1000	10000	1.27	1.41
.	.	.	.	...	.	.	.	.	.
n	0	1	0	...	0	500	1500	−0.09	−0.65

view data set of nearly 10,000 unique views. For each, we collected the prefix, the population range, and the total number of occupations.

One can distinguish between textual and numerical attributes. Textual attributes correspond to view prefixes, whereas numerical attributes to population ranges. As shown in Table 1, each view is defined by a collection of binary-valued attributes, namely a, b, c, ..., z, where one of them is set to one—meaning that the view was produced using that prefix—and the rest to zero. Numerical attributes, on the other hand, are represented by the minimum and maximum numbers characterizing a population range. For the example below, we assume that the average and standard deviation for the attribute *min* are 533.33 and 368.18, respectively, and for the attribute *max* 4166.67 and 4129.84. Finally, we take advantage of this off-line computation to obtain the zscore of each numerical attribute which is used in the on-line recommendation process described in the next section.

As a means to expose potential views of interest, our system uses the seen views in the view data set to reveal views not seen yet by the user but still adhering to his/her user profile. This approach is comparable to the "discover the unexpected" as coined by Thomas [22]. Our method is based on the premise that preferences for a view is related to preferences for properties of that view. It uses only the current user's navigational data in conjunction with view property data to make recommendations.

4.2 User Profile

In the second component, we first calculate the preferences for seen views, then the preferences for attributes of such views. In the former, we use a method based on the dwell time as an implicit interest indicator; in the latter, the preferences of attributes are obtained for both numerical and textual attributes.

A user profile $P \subseteq V$ is defined as a collection of seen views, i.e., views produced by the user while interacting with the system.

Definition 1   We defined the set of unseen views U as the complement of P with respect to V

$$U = V \setminus P = \{x | x \in V \wedge x \notin P\} \tag{1}$$

4.2.1 Inferring Preferences for Seen Views

Definition 2   The set of preferences for seen views v in P is defined as

$$\{PRF_v = z(t_v)/\varepsilon | v = 1, \ldots, |P|\} \tag{2}$$

where $z(t_v)$ is the trimmed zscore (cf. Algorithm 1) of the dwell time t of view v, and ε a threshold for outliers account and normalization. Following the recipe in [16], we set ε to 3.

Algorithm 1 Compute preferences for views

---

   procedure VIEWPREF(P)
      $a \leftarrow$ average of all view dwell times in P
      $s \leftarrow$ standard deviation of all view dwell times in P
      for $v=1$ to $|P|$ **do**
         $z(t_v) \leftarrow \frac{t_v - a}{s}$ // zscore of dwell time t of view v
         if $z(t_v) > \varepsilon$ **then**   // trimming zscore
            $z(t_v) \leftarrow \varepsilon$
         else if $z(t_v) < -\varepsilon$ **then**   // trimming zscore
            $z(t_v) \leftarrow -\varepsilon$
         end if
         $PRF_v \leftarrow z(t_v)/\varepsilon$ // preference for view v
      end for
      return $\{PRFv\}$ // a set of view preferences
   end procedure

4.2.2 Inferring Preferences for Attributes of Seen Views

The preferences of attributes are obtained for both numerical and textual attributes (see Algorithm 2). The output of the second component is a pair of sets as defined in the following.

Definition 3   The set of preferences for textual attributes x in α is defined as the weighted mean of view preferences in all seen views.

$$\{PRF_{\alpha_x} = \frac{\sum \alpha_x(v) PRF_v}{\sum \alpha_x(v)} | x = 1, \ldots, |\alpha|\} \tag{3}$$

where $\alpha_x(v)$ is the value of textual attribute x of view v in P.

Definition 4 The set of preferences for numerical attributes x in β is defined as the weighted mean of its values in all positive samples, i.e., all seen views with positive view preference values, as defined in (2).

$$\{PRF_{\beta_x} = \frac{\sum\limits_{PRF_v > 0} PRF_v \beta_x(v)}{\sum\limits_{PRF_v > 0} PRF_v} \, | x = 1, \ldots, |\beta|\} \tag{4}$$

where $\beta_x(v)$ is the value of numerical attribute x of view v in P.

Algorithm 2 Compute preferences for attributes

---

procedure ATTPREF($\{PRF_v\}$)
    for x=1 to $|\alpha|$ **do**   // preferences for textual attributes
        calculate PRF_{α_x}
    end for
    for x=1 to $|\beta|$ **do**   // preferences for numerical attributes
        calculate PRF_{β_x}
    end for
    return $\{PRF_\alpha\}, \{PRF_\beta\}$ // two sets: textual-attribute preferences, and numerical-attribute preferences
end procedure

4.3 Content-Based Recommendation

Finally, we obtain the degree of desirability for all views in the view data set. We first calculate the similarity between attribute values in views and the attribute preferences. Formally, the set of similarity of textual attributes in view v with respect to the textual attribute preferences is defined by

$$\{S_{\alpha_x(v)} = \frac{1 + f(\alpha_x(v))}{2} \, | x = 1, \ldots, |\alpha|\} \tag{5}$$

where $\alpha_x(v)$ is the value of textual attribute x of view v in V, and the function f is defined by the rule

$$f(\alpha_x(v)) = \begin{cases} PRF_{\alpha_x} & \text{if } \alpha_x(v) = 1 \\ 0 & \text{otherwise} \end{cases} \tag{6}$$

Likewise, the set of similarity of numerical attributes in view v with respect to the numerical attribute preferences is defined by

$$\{S_{\beta_x(v)} = 1 - \frac{|z(PRF_{\beta_x}) - z(\beta_x(v))|}{2\varepsilon_{\beta_x}} \, | x = 1, , |\beta|\} \tag{7}$$

where $z(x)$ is the trimmed zscore of x calculated based on the mean and standard deviation of the corresponding numerical attribute for all views in V. $\beta_x(v)$ is the value of numerical attribute x of view v in V, and ε_{β_x} is a threshold of numerical attribute x.

Finally, S_α and S_β are combined to produce a single index that represents the degree of desirability for each view in the view data set. Since some attributes are more important than others, the desirability for view v is a weighted mean defined by

$$D_v = \frac{\sum R_{\alpha_x} S_{\alpha_x}(v) + \sum R_{\beta x} S_{\beta_x}(v)}{\sum R_{\alpha_x} + \sum R_{\beta_x}} \tag{8}$$

where R_{α_x} and R_{β_x} denote the relative weights for textual and numerical attributes x, respectively.

The system then recommends the top-n unseen views in U with the highest desirability.

4.4 Usage Scenario

We now want to give a step-by-step demonstration of how a typical visual-interaction session takes place. The goal is to highlight the major features employed by our system to recommend views not seen by the user, but still adhering to his/her user profile. We illustrate this using a sequence of two different recommendation sessions. On each session, the system produces the top-5 recommendation set of views, sorted from highest to lowest rank number.

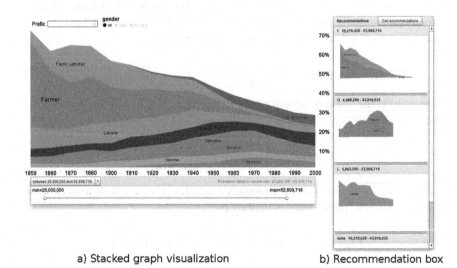

a) Stacked graph visualization   b) Recommendation box

Fig. 4 First recommendation session from initial view and dwell time of 10 s

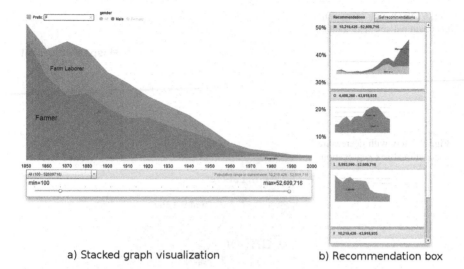

a) Stacked graph visualization   b) Recommendation box

Fig. 5 Second recommendation session, with preference on textual attribute, and dwell time of 80 s

The initial view shown in Fig. 4a illustrates a view containing one of the initial attribute settings employed in our system evaluation: *prefix* = none, *min* = 25,000,000 and *max* = 52,609,716. Figure 4b shows the results of the first top-5 recommendation session (only three views are shown here due to limited space) from a dwell time of 10 s. A strong similarity in the value distribution of *min* and *max* can be observed between the initial and the recommended views. The system also tries to diversify the values for the *prefix* attribute. Furthermore, the initial view, already seen by the user, is not recommended by the system. Soon after, the top-1 view is selected from the recommendation list. As shown in Fig. 5a, this view becomes the current one in the system and, as before, will not be recommended in subsequent recommendation sessions. The property values of this view are as follows: *prefix* = F, *min* = 10,219,426, *max* = 52,609,716. Finally, in the second recommendation session, Fig. 5b, from a dwell time of 80 s, the system recommended a view with *prefix* = F, which shows the effects of a a higher dwell time on the selected view. Also, the system keeps similar value distribution of occurrence frequencies.

5 User Study

Using an approach similar to that used in [16], we conducted a usability study. Our goal was to explore the quality of the recommendations produced by our system. For this, we used a comparison page containing high and low ranked views, and we expected the participants to show interest in the high-ranked-view group. This method is described later in this section.

Fig. 6 View with increasing
occupational sex segregation

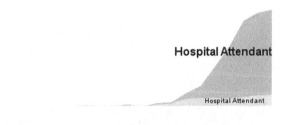

Fig. 7 View with decreasing
occupational sex segregation

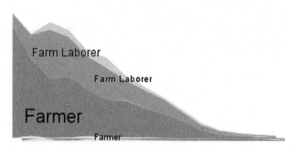

We employed a concept known as *Occupational Sex Segregation*, which refers
to a sociological term [23, 24] used for specific occupations in which women or
men are over- or underrepresented, i.e., they work almost exclusively with their
same gender. One can distinguish between increasing and decreasing occupational
sex segregation. Increasing segregation refers to occupations in which at the
beginning of the period a weak segregation can be observed, though later this
segregation is increased. Conversely, decreasing segregation refers to occupations
with strong segregation at the beginning and decreasing segregation later in time.
In Fig. 6, Hospital Attendant shows increasing segregation, while Farm occupations, in Fig. 7, decreasing segregation. In both cases the population worked
almost exclusively with their same gender.

We recruited 32 participants to participate in a visual analytics activity.
Participants in this study were computer science students taking courses at our
university and consisting of 21 undergraduate students and 11 graduate students
with the average age of 22.3, ranging from 20 to 27 years old. We hypothesized
that recommendations based on rankings calculated by our system were better than
randomly generated recommendations.

We prepared two tasks involved in searching, using the features of the visualization system, for occupations the participants believed had evidences of sex
segregation. To begin with, each participant was given an introductory tutorial to
Occupational Sex Segregation as well as the user guide of the visualization system. Next, they were asked to use the system for 10 min. The idea behind this is
that participants could learn how to use the system before doing the actual tasks.
After this period, they were given two population ranges: 10,000,000–25,000,000
and 25,000,000–52,609,716. Then, for each one, they had to answer the following
question—*which is the most segregated occupation?* For each population range,
each participant had 3 min to provide an answer.

While each participant was looking for segregated occupations, the system collected both dwell time and properties for all produced views. After three minutes, all unseen views in the view data set were ranked and a comparison page was generated for use in the subsequent phase. The purpose of the comparison page was to display unseen views that were ranked high and low by the system, and measure how frequently participants chose high ranked views on the comparison page. If the system produces good rankings based on preferences, participants should pick high ranked views most of the time. Thus, this evaluation provides a relatively simple yet reliable way of determining how well the system works.

For each participant, the comparison page consisted of 12 unseen views, with 6 views being high ranked and 6 views being low ranked. These views were obtained from the top-100 unseen views in the view data set. From them, the high ranked views were selected from the top-6, and the low ranked views from the bottom-6. To prevent biases that might arise from position effects, the positions of those 12 unseen views were randomly decided. The participant was instructed to choose five views, in terms of their usefulness for supporting the answer. After viewing the comparison page, the participant was routed to the last page in which, from the five selected views, he or she had to select the most-like view. Finally, the participant was asked to write the reasons that motivated the second selection.

6 Results and Discussions

In this section, we report observations from our study on user profile generation. The data analyzed were drawn from usage logs including 320 views chosen by the participants in the first selection, and 64 views chosen in the second selection. We analyzed the results under the premise that the recommendations provided by our system are correct if users can find their answers, and if those answers are supported by our system.

We first wanted to know the frequency that high ranked views were chosen by the participants during the first selection. From the second selection, we wanted to know the rank of chosen views and whether those views were useful for the participants. Are they high ranked views?, do they contain the users' answers? Finally, we wanted to learn from these participants their reasons that motivated the second selection.

From the first selection, we computed the percentage of times that participants chose views that were ranked high by the system. If view rankings were random, we would expect 50 % of the higher ranked views be chosen in the first selection. However, those views were chosen by participants 63.75 % of the time. This difference is statistically significant (p-value: 0.01) according to the Chi-square test. Although a 63.75 % effectiveness score does not seem high, this should be viewed in the context that the system recommended only unseen views.

From the second selection, we computed the percentage of times that the participants chose a high ranked view. Since the selection space was produced by each participant from the first selection we expected a high percentage here. Views in the high ranked group were chosen 68.75 % of the time. This difference, from 63.75 % is not statistically significant (p-value: 0.05) and shows a consistent user criteria between the two selections.

7 Conclusion and Future Work

In this chapter, we raised the question of how recommending can be incorporated into an information visualization system for suggesting interesting views to the user. Aiming to alleviate the overload caused in the visual exploration of more than 10,000 views, we proposed a method for adapting a recommender engine to the stacked graph visualization. Our method employs a content-based recommendation algorithm which uses dwell times to infer user preferences on stacked graph views. We rely on the premise that preferences for a view is related to preferences for properties of that view. A usability study with more than 30 subjects shows that (1) accurate user profile can be generated by analyzing user behavior and view property data, (2) dwell times can be a useful indicator of preferences, (3) preference for a view can be expressed as a preference function of view-attribute values, and (4) attribute preferences can be transferred across views.

While our system shows promise, there remain several areas that can be improved upon. We would like, for example, to explore alternative interest indicators for inferring user preferences such as mouse movements, annotations on a view, revisits on a view, etc. In our previous work, we tested varios combinations of them and would like to know their effectiveness when applied to the system.

References

1. Konstan, J.A.: Introduction to recommender systems: algorithms and evaluation. ACM Trans. Inf. Syst. **22**(1), 1–4 (2004). http://doi.acm.org/10.1145/963770.963771
2. Ricci, F., Rokach, L., Shapira, B.: Introduction to recommender systems handbook. Recommender Systems Handbook, pp. 1–35. Springer, Boston (2011)
3. Claypool, M., Le, P., Wased, M., Brown, D.: Implicit interest indicators. In: Proceedings of the 6th International Conference on Intelligent User Interfaces IUI '01, pp. 33–40. ACM, New York (2001)
4. Kelly, D., Teevan, J.: Implicit feedback for inferring user preference: a bibliography. SIGIR Forum **37**(2), 18–28 (2003)
5. Morita, M., Shinoda, Y.: Information filtering based on user behavior analysis and best match text retrieval. In: Proceedings of the 17th Annual International ACM SIGIR Conference on Research and Development in Information Retrieval, ser. SIGIR '94, pp. 272–281. Springer-Verlag New York, Inc, New York (1994)
6. Kelly, D.Teevan, J.: Implicit feedback for inferring user preference: a bibliography. SIGIR Forum **37**(2), 18–28 (2003). http://doi.acm.org/10.1145/959258.959260

7. Riche, N., Lee, B., Plaisant, C.: Understanding interactive legends: a comparative evaluation with standard widgets. Comput. Graph. Forum **29**(3), 1193–1202 (2010). (Wiley Online Library)
8. Viegas, F.B., Wattenberg, M., van Ham, F., Kriss, J., McKeon, M.: Manyeyes: a site for visualization at internet scale. IEEE Trans. Vis. Comput. Graph. **13**, 1121–1128 (2007)
9. Willett, W., Heer, J., Agrawala, M.: Scented widgets: improving navigation cues with embedded visualizations. IEEE Trans. Vis. Comput. Graph. **13**, 1129–1136 (2007)
10. Vuillemot, R., Clement, T., Plaisant, C., Kumar, A.: What's being said near "Martha" ? Exploring name entities in literary text collections. In: IEEE Symposium on Visual Analytics Science and Technology, VAST 2009, pp. 107–114 (2009)
11. Crossno, P., Dunlavy, D., Shead, T.: LSAView: a tool for visual exploration of latent semantic modeling. In: IEEE Symposium on Visual Analytics Science and Technology, VAST 2009, pp. 83–90 (2009)
12. Koop, D.: Viscomplete: automating suggestions for visualization pipelines. IEEE Trans. Vis. Comput. Graph. **14**, 1691–1698 (2008)
13. Crnovrsanin, T., Liao, I., Wu, Y., Ma, K.: Visual recommendations for network navigation. Comput. Graph. Forum **30**(3), 1081–1090 (2001). (Wiley Online Liabrary)
14. Gotz, D., Zhou, M., Aggarwal, V.: Interactive visual synthesis of analytic knowledge. In: Proceedings of the IEEE Symposium on Visual Analytics Science and Technology, pp. 51–58 (2006)
15. Yang, D., Rundensteiner, E., Ward, M.: Analysis guided visual exploration of multivariate data. In: Proceedings of the IEEE Symposium on Visual Analytics Science and Technology, Citeseer, pp. 83–90 (2007)
16. Parsons, J., Ralph, P., Gallagher, K.: Using viewing time to infer user preference in recommender systems. In: Proceedings of the AAAI Workshop on Semantic Web Personalization. American Association for Artificial Intelligence, pp. 52–63 (2004)
17. Heer, J., Viégas, F.B., Wattenberg, M.: Voyagers and voyeurs: supporting asynchronous collaborative visualization. Commun. ACM **52**(1), 87–97 (2009)
18. Wattenberg, M., Kriss, J.: Designing for social data analysis. IEEE Trans. Vis. Comput. Graph. **12**(4), 549–557 (2006)
19. Havre, S., Hetzler, E., Whitney, P., Nowell, L.: Themeriver: visualizing thematic changes in large document collections. IEEE Trans. Vis. Comput. Graph. **8**, 9–20 (January 2002)
20. Byron, L., Wattenberg, M.: Stacked graphs—geometry & aesthetics. IEEE Trans. Vis. Comput. Graph. **14**, 1245–1252 (2008)
21. Heer, J., Kong, N., Agrawala, M.: Sizing the horizon: the effects of chart size and layering on the graphical perception of time series visualizations. In: Proceedings of the 27th International Conference on Human Factors in Computing Systems, pp. 1303–1312. ACM (2009)
22. Thomas, J.J., Cook, K.A. (eds.): Illuminating the path: the research and development agenda for visual analytics. IEEE Computer Society Press
23. Jacobs, J.: Long-term trends in occupational segregation by sex. Am. J. Sociol. **95**(1), 160–173 (1989)
24. Charles, M., Grusky, D.: Occupational ghettos: the worldwide segregation of women and men. Stanford University Press, Stanford (2005)

Pattern Extraction from Graphs and Beyond

Hiroshi Sakamoto and Tetsuji Kuboyama

Abstract We explain recent studies on pattern extraction from large-scale graphs. Patterns are represented explicitly and implicitly. Explicit patterns are concrete subgraphs defined in graph theory, e.g., clique and tree. For an explicit model of patterns, we introduce notable fast algorithms for finding all frequent patterns. We also confirm that these problems are closely related to traditional problems in data mining. On the other hand, implicit patterns are defined by statistical factors, e.g., modularity, betweenness, and flow determining optimal hidden subgraphs. For both models, we give an introductory survey focusing on notable pattern extraction algorithms.

1 Introduction

We review recent research on pattern extraction from large-scale networks to provide a guideline for beginners. A network is a set of links connecting objects, which is equivalent to a graph in discrete mathematics. In this survey, we usually use *graphs* instead of *networks*, because graph theory contains networks as a different notion. Here, not all types of networks are attractive. We particularly pay

Partially supported by KAKENHI(23680016, 20589824) and JST PRESTO program.

H. Sakamoto
Kyushu Institute of Technology, 680-4 Kawazu, Iizuka-shi, Fukuoka 820-8502, Japan

T. Kuboyama
Gakushuin University, 1-5-1 Mejiro, Toshima-ku, Tokyo 171-8588, Japan
e-mail: kuboyama@tk.cc.gakushuin.ac.jp

H. Sakamoto (✉)
PRESTO JST, 4-1-8 Honcho, Kawaguchi, Saitama 332-0012, Japan
e-mail: hiroshi@ai.kyutech.ac.jp

G. A. Tsihrintzis et al. (eds.), *Multimedia Services in Intelligent Environments*,
Smart Innovation, Systems and Technologies 24, DOI: 10.1007/978-3-319-00372-6_7,
© Springer International Publishing Switzerland 2013

attention to heterogeneous networks. For example, consider a random graph obtained by connecting any pair of objects under uniform probability. Such a network is almost homogeneous and thus it is unlikely that an interesting pattern will be found. In contracts, we have many examples of large-scale real-world networks, such as friendship networks and WWW link structures; they are heterogeneous. For example, a blogging community contains a network of declared friendships defined by over 4.4-million-nodes [7], and a network of all communications over a month on Microsoft Instant Messenger has over 240-million-nodes [39]. Our objective is to find patterns and statistical properties of such networks using a mathematical model and algorithms to analyze graphs.

Various clustering and community detection methods have been proposed in the last decade, e.g., *hierarchical*, *graph partitioning*, *spectral*, *clique percolating*, etc. Although we cannot catalog all the studies, we focus on the aspect of large-scale data in this survey.

Girvan and Newman [28] defined a property of community structure, in which network nodes are tightly joined to each other inside groups, between which there are only looser connections. They proposed a method for detecting such communities based on *centrality* indices to find community boundaries. Following Girvan and Newman's work, many algorithms for this problem have been proposed. Newman and Girvan [51] introduced *edge betweenness* and *modularity* to evaluate the adequacy of current communities. The edge betweenness is the number of shortest paths[1] through the edge in question. The modularity is, intuitively, the difference of the number of edges and expected number of edges within communities. Because computing both betweenness and modularity is expensive, even if assuming sparse graphs, Newman [50] proposed a faster algorithm using modularity by only approximating Newman and Girvan's algorithm. Since then, many improvements have been proposed; e.g., [10, 17].

Another important community detection is based on graph partitioning to minimize the number of within-community connections. This is approximately equivalent to minimize the number of between-community connections. There are many approaches to solve it, e.g., spectral graph partitioning. We, however, focus on another method based on connectivity and transitivity in large-scale graphs. Flake et al. [23] defined the dense substructure using graph terminology: a community, i.e., a dense substructure, is a set of nodes in which each member has at least as many edges connecting to members as it does to non-members. They also proposed the method to extract communities from large-scale graphs based on a traditional graph algorithm. Inspired by the Flake study, the authors applied the method to meta-network, which is implicitly represented by references in research papers, and extracted research communities, where it is difficult to find such communities by the usual keyword search [48].

---

[1] There are other measures to define betweenness instead of shortest path. The shortest path is, however, easier to compute than other measures.

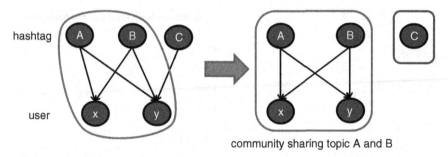

Fig. 1 Community as maximal clique in bipartite graph

Another idea for representing interesting patterns derives from frequency. In the last decade, we have been able to access huge networks in the Web and SNSs. This situation makes the extraction problem difficult using conventional algorithms and data structures. To address the problem, we next focus on fast and space-saving algorithms. For this problem, Makino and Uno [43] developed an efficient algorithm to enumerate all maximal *cliques*, i.e., a set of nodes embedded as a complete subgraph. This algorithm is the fastest that achieves *polynomial time delay*; i.e., the time to output a clique is bounded by a polynomial in input size. For this enumeration problem, Tomita et al. [66] developed an optimal algorithm for the total running time using graph theory technique. Uno [68] proposed an algorithm to enumerate frequent *closed* itemsets in data mining from transaction data. Using the well-known fact that the closed itemset in transaction data is identical to the maximal bipartite clique, this algorithm solved the problem of enumerating all maximal bipartite cliques in an input bipartite clique with polynomial time delay. Figure 1 shows an example of maximal bipartite clique as a community in a bipartite graph representing the relationship between users and hashtags in the tweets they used. As we see in this example, note that any network represented by a graph can be transformed to an equivalent bipartite graph; thus, Makino and Uno's algorithm and Uno's algorithm are sufficiently general.

On a related study, we have addressed the ranking problem of frequent *tweets* from a huge number of the twitter texts. The collection of tweets contains many hash tags associated with each other; it is regraded as a meta-network. Figure 2 displays the number of frequent hashtags in twitter data from 11 March to 29 March. The data size, however, is huge; e.g., the tweets for the 311 earthquake in Japan, including #tsunami or #jishin (earthquake in Japanese), is greater than 100GB of data in the Japanese language. For this difficulty, the authors applied Uno's enumeration algorithm to the hidden network data for detecting communities. Figure 2 also shows the number of extracted communities by Uno's enumeration algorithm.

The algorithms mentioned above are categorized into two classes: explicit and implicit models. Explicit patterns are concrete subgraphs; e.g., clique and tree, and implicit patterns are also subgraphs defined by statistical factors: modularity

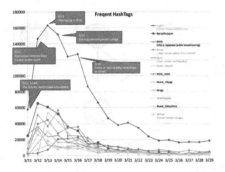 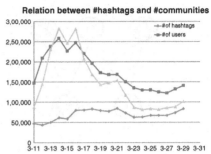

Fig. 2 The number of frequent hashtags in tweets for the Great East Japan Earthquake (*left*) and the number of extracted frequent maximal cliques from the tweets (*right*)

and flow for example. For both models, we give an introductory survey on notable pattern extraction algorithms.

Finally, we focus on the next stage: the time series network data. Indeed, most of the big data, including the example mentioned above, contains time attribution. In the last part of our survey, we summarize recent challenges in this research field. In analyzing time series networks, most studies try to find structural relationships among snapshot community structures over a time series network after extracting communities in each snapshot. This problem addresses two sub-problems: graph partitioning and historical mapping among time series networks. We review existing methods from the standpoint of the two sub-problems.

2 Foundations

This section presents basic concepts used later in this chapter. Section 2.1 consists of standard notions in graph theory related to community extraction. For further details on graph theory, see [21]. Section 2.2 summarizes how to compactly represent a graph, a mathematical object, using a computer. Section 2.3 introduces the graph algorithms on which many community extractions are based. For further details of these data structures and algorithms, see [19].

2.1 Graphs

For a set S, S^k denotes the set of all k-element subsets of S. $|S|$ is the number of cardinality of S. Assume finite sets V and $E \subseteq V^2$. An (undirected) graph is a pair (V, E), where $u \in V$ (resp. $e = \{u, v\} \in E$) is called a *vertex* or *node* (resp. *edge*). Especially, an edge $\{u, u\}$ is called a *loop*. A *directed* graph is a graph (V, E) together with maps $start : E \to V$ and $end : E \to V$ such that for each

$e = \{u, v\} \in E$, $start(e) \in \{u, v\}$ and $end(e) = \{u, v\} - start(e)$, i.e., any edge $\{u, v\}$ is oriented as either (u, v) or (v, u). The set of nodes of a graph G is referred to as $V(G)$, its set of edges as $E(G)$. $|V(G)|$ and $|E(G)|$ are called the *order* and *size* of G, respectively.

For $u, v \in V(G)$, they are called *adjacent* if $\{u, v\} \in E(G)$ and called *independent* otherwise. This notion is also used for edges.

The number of different nodes adjacent to a node v, denoted by $d(v)$, is called the *degree* of v. A node with degree 0 is *isolated*. $\delta(G)$ is the minimum degree and $\Delta(G)$ is the maximum degree of $v \in V(G)$.

A path is a graph $P = (V, E)$ such that $V = \{x_1, x_2 \ldots, x_n\}$ and $E = \{\{x_1, x_2\}, \{x_2, x_3\}, \ldots, \{x_{n-1}, x_n\}\}$, where all x_i are different. $|E|$ is called the *length* of the path. If $n - 1 \geq 3$ and $x_n = x_1$, the path P is called a *cycle*. A graph not containing any cycle is *acyclic*. A graph is called *connected* if there is at least one path between any two nodes of it.

If any two different nodes in $V(G)$ are adjacent, G is called a *complete* graph. If $|V(G)| = n$, the complete graph is written K_n. If $V(G) = V_1 \cup V_2$ and $V_1 \cap V_2 = \emptyset$ such that any $\{u, v\} \in E(G)$ satisfies $u, v \in V_1$ or $u, v \in V_2$, G is called *bipartite* graph. A connected acyclic graph T is called a *tree*. A node with degree 1 is a *leaf*. When a node in T is fixed as a special node called the *root*, T is a rooted tree. The nodes other than leaves are called *internal* nodes. There is exactly one path from the root to any node v. The path length is the depth of v, denoted by $depth(v)$. If $depth(v) \leq depth(u)$, v is an *ancestor* of u and u is a *descendant* of v. Specifically, if $depth(v) = depth(u) + 1$, v is a *child* of u and u is the *parent* of v. When the children of any node v are comparable, the tree is *ordered*. G is called a *directed* graph (*digraph*) if any edge $\{u, v\} \in E(G)$ is oriented as (u, v) or (v, u) (possibly both of them). We write edges of digraphs with the orientation. A directed acyclic graph is denoted by DAG.

$G' = (V', E')$ is called a subgraph of G, denoted by $G' \subseteq G$, if $V' \subseteq V$ and $E' \subseteq E$. If $G' \subseteq G$ and $\{u, v\} \in E(G)$ implies $\{u, v\} \in E'(G)$ for all $u, v \in V'$, G' is called an *induced* subgraph of G. A subset of $V' \subseteq V(G)$ of size k is called *k-clique* if any two nodes in V' are adjacent. A subgraph T of G is a *spanning tree* if T is a tree and $V(T) = V(G)$. Subgraphs and related notions are illustrated in Fig. 3. In addition, we give the first three non-trivial complete graphs and a complete bipartite graph $K_{3,2}$ together with the corresponding cliques embedded into a graph (left) and bipartite graph (right) in Fig. 4.

2.2 Graph Representations

A graph is a good model to represent structures that mean a kind of dependency, e.g., information network and social network. As an information network, we can find a huge digraph in the World Wide Web, where two pages correspond to nodes u, v and a link form u to v is represented by the edge (u, v). We can also find such a digraph in the implicit network like the relationship between researchers and their

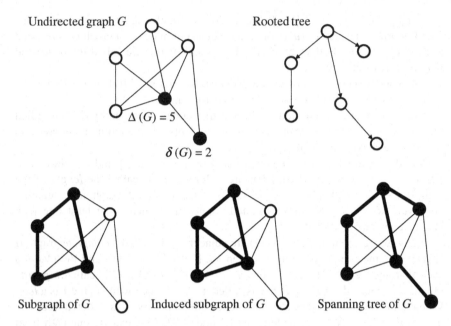

Undirected graph G Rooted tree

$\Delta(G) = 5$

$\delta(G) = 2$

Subgraph of G Induced subgraph of G Spanning tree of G

Fig. 3 Embedded subgraphs

interests. On the other hand, as an example of social networks, we can define a bipartite graph $G = (V_1 \cup V_2, E)$ for a person $u \in V_1$ and an affiliation $v \in V_2$ such that u belongs to v iff $\{u, v\} \in E$. To handle such a large graph by computer, we need a compact and directly accessible representation. We introduce tree types of representation models for developing graphs. There is trade-off between time and space complexities dependent on the models. The size of adjacency list is smaller; however, the time to access is slower than with an adjacency matrix (Fig. 5).

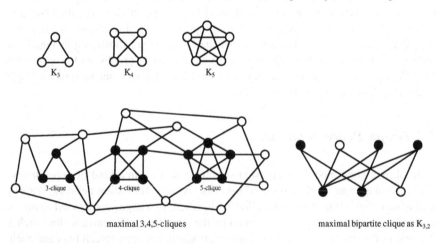

K_3 K_4 K_5

3-clique   4-clique   5-clique

maximal 3,4,5-cliques   maximal bipartite clique as $K_{3,2}$

Fig. 4 Complete graphs and maximal cliques

Adjacency list: For a graph $G = (V, E)$ with $|V| = n$, its adjacency list is defined by an array $A[1, n]$ and at most n lists L_i such that each $A[i]$ indicates the head of L_i consisting the nodes adjacent to $v_i \in V$. Thus, each edge $\{v_i, v_j\} \in E$ appears twice as a member of L_i and L_j. When $|V| = n$ and $|E| = m$, this data structure requires at most $O(n + m)$ space. This space complexity is much smaller than that of an adjacency matrix in the worst case. The list representation, however, requires $O(\Delta(G))$ time to check if $\{u, v\} \in E$. This existence check can be performed in $O(1)$ time using the matrix model below. Figure 5 shows an example of the adjacency list/matrix representation of a graph. An important application of the adjacency list is the tree representation. We see two trees in Fig. 6: one is unbounded with the degree and the other is bounded (in-degree is at most one and out-degree is at most 2). Of course, they are not equivalent; we can adopt the later alternative to the former to traverse the tree. Beyond that, if we add the pointers from each node to its parent, the expression is equivalent to the original tree.

Adjacency matrix: An adjacency matrix is a $0, 1$-matrix $A[1, n][1, n]$ such that $A[i][j] = 1$ if node v_i adjacent to v_j and $A[i][j] = 0$ otherwise. A is symmetrical for undirected graphs. With digraphs, $A[i][j] = 1$ iff $(v_i, v_j) \in E$. In this model, we can access any edge in $O(1)$ time. On the other hand, the space complexity is $\Theta(n^2)$. When a graph is sparse so that m is much smaller than n, $O(n + m)$ is the desirable space for our model. Indeed, WWW and a social network are usually sparse graphs.

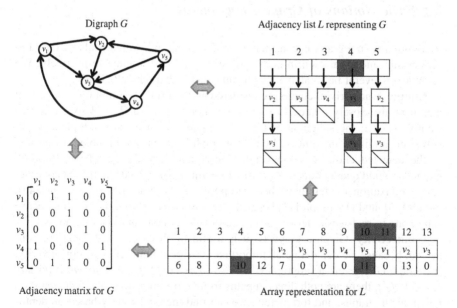

Fig. 5 Graph representation by list and matrix

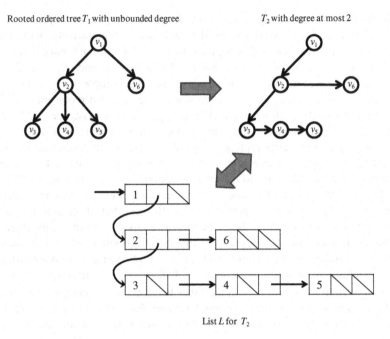

Fig. 6 Simulation of multi-branching tree by list

2.3 Basic Notions of Graph Components

A community in a graph is considered a dense subgraph, which is also referred to as a cluster and a partition in graph clustering and partitioning, respectively. We summarize several basic algorithms to find such subgraphs efficiently.

Clique: A non-trivial clique is considered an important community in which all members are closely related to each other. For any subset $V' \subset V$, we can efficiently check if the subgraph of $G = (V, E)$ induced by V' is a complete graph. However, finding a maximum clique in a graph is an NP-hard problem if the size of the detected clique is unbounded. Consequently, researchers have proposed methods to find quasi-cliques or maximal k-cliques for a fixed small k. Many other clique relaxation models have been proposed, such as k-plex [61], n-clan [45], n-club [45], and DN-graph [71] because strict cliques are too tight to extract dense subgraphs in real-world networks, in addition to its computational hardness.

Connectivity: Given a graph G, it is useful to check if $u, v \in V(G)$ are connected, because we can define an important class of communities by connectivity. A digraph G is given, a subgraph G' is a *strongly connected component* if for any $u, v \in V(G')$, there are both directed paths from u to v and v to u. We focus on such maximal subgraphs, and fortunately, we can find them efficiently based on depth-first search on digraphs.

Fig. 7 Relation between maximum flow and minimum cut of network

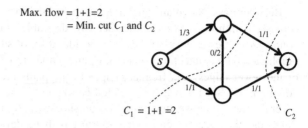

Max. flow = 1+1=2 = Min. cut C_1 and C_2

$C_1 = 1+1 = 2$

C_2

***k*-core and *k*-truss**: A k-core [60] and k-truss [18] are also regarded as clique relaxation models. A k-core is a maximal connected subgraph in which every node has a degree of at least k. The k-cores in a graph are obtained by repeatedly removing all nodes with a degree less than k until every remaining node has a degree of at least k. A graph is decomposed into a hierarchical structure of k-cores.

Network flow and cut: The above three notions require a condition that must be satisfied by all nodes in an extracted subgraph. Network flow and cut define a weaker condition for communities to be extracted. Let G be a digraph. Assume two nodes $s, t \in V(G)$ such that any node is reachable from s, and t is reachable from any node. A subset $C \subseteq E$ is called a *cut* of G if t is not reachable in the graph $(V(G), E(G) - C)$. A minimum cut that minimizes $\sum_{e \in C} w(e)^2$ is naturally associated with communities. A cut divides the inside community including s from the outside including t. By the max-flow min-cut theorem, we expect that a minimum cut maximizes the density of extracted nodes by a maximum flow, and a max-flow can be found efficiently. A relation between max. flow and min. cut is shown in Fig. 7. In this figure, each edge is labeled f/c, where f is the size of flow running through the directed edge and c is the capacity of the edge. Thus, the current flow of this network is the sum of fs from the source, which is equal to the sum of fs to the sink. On the other hand, there are two different min. cuts: the sum of capacities of directed edges from V_s to V_t. They are shown by broken lines.

Enumeration: Enumeration is a problem of outputting all elements in a class, where no element is counted more than once. This technique is useful for data mining and graph mining, because enumeration of frequent itemsets in transaction data is an important problem in data mining; it is equivalent to the enumeration of cliques in a bipartite graph. Finding the maximum clique is NP-hard, but finding maximal cliques is tractable. Indeed, there are efficient algorithms to solve this problem. However, the number of outputs is potentially exponential. To evaluate an enumeration algorithm, *polynomial time delay* is often adopted; the time to output a next element is bounded by a polynomial in the input size. The difficulty of the enumeration problem depends on the class of objects. Other than maximal cliques, we can effectively enumerate the ordered and unordered trees that are related to graph mining problem.

---

[2] Assume any edge has unit weight in web graph.

Betweenness: Newman and Girvan proposed the following idea to decide edges across communities: We assume that two desirable communities C_i and C_j are connected by fewer edges. When we consider a set of sufficiently many paths in a graph, any path connecting a node in C_i and a node in C_j must contain at least one of the fewer edges; i.e., frequent edges in the paths are expected to connect the communities. Such a measure is called *betweenness*. Newman and Girvan defined several concrete betweennesses based on shortest path, random walk, and current flow, and they recommended the shortest path model is most efficient through complexity and experiments. This measure defines which is expected to be an edge between communities, however, we cannot decide the communities themselves by betweenness only. To measure edge density inside communities, Newman and Girvan also proposed another notion given below.

Modularity and its approximation: *Modularity*, a quality function, to measure randomness of a graph G was proposed by Newman and Girvan. The idea is that a random graph is not expected to contain heterogeneous structures. In other words, desirable community C is one in which edges inside C are dense and edges between C and other communities are sparse. Newman and Girvan defined the modularity Q by

$$Q = \sum_{c=1}^{n_c} \left\{ \frac{e_c}{m} - \left(\frac{d_c}{2m} \right)^2 \right\},$$

where c is the identifier of a community, n_c is the number of communities, m is the number of edges, and e_c is the size of community c; i.e., the number of edges inside of c, and d_c is the sum of $d(v)$ for any node v in c. In this definition, the term e_c/m is the edge density of community c. If we define a modularity by $Q' = \sum_{c=1}^{n_c} \frac{e_c}{m}$, trivially, the single community that coincides with the whole graph maximizes Q'. To avoid such inexpedience, we need the second term, which means the number of all edges adjacent to nodes in c. By this definition, $0 \le Q \le 1$ and Q comes to 0 when c is a random graph. Thus, we can estimate the density of current communities.

3 Explicit Models

Because the *subgraph isomorphism* is NP-hard, we need to restrict the class of extracted patterns in a practical sense. Thus, we consider the method to extract simple implicit patterns from graphs, i.e., trees, cliques, and other local substructures. For trees, we treat both ordered and unordered versions. For cliques, we basically consider maximal k-cliques, because finding *maximum* clique is computationally hard. These problems of finding partial structure from a large graph are fundamental to pattern discovery and data mining. Indeed, the problem of finding maximal cliques on a bipartite graph is equivalent to the problem of finding frequent closed itemsets from transactions.

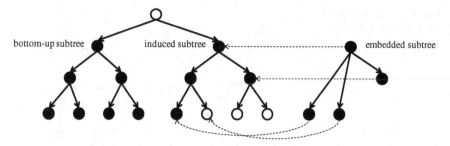

Fig. 8 Bottom-up, induced, and embedded subtrees

3.1 Tree

Tree patterns: We consider the two classes of rooted trees, *ordered and unor-dered*.[3] Again, a tree is ordered if any two children are comparable, and unordered if such a binary relation is undefined. Given a tree $T = (V, E)$, its subgraph $T' = (V', E')$ with root v is called a *bottom-up subtree* if T' is a connected subgraph of T such that T' contains all descendants of v in T, called *induced subtree* if T' is a connected induced subgraph of T, and called *embedded subtree* if there exists a one-to-one mapping from V to V' preserving the ancestor-descendant relationship of T. For these subtree patterns, we assume that the labels and order among siblings (in case of ordered trees) are also preserved. We can find a trivial inclusive relationship: bottom-up subtree $\subset$ induced subtree $\subset$ embedded subtree (Fig. 8). Although other subtree patterns have been proposed, we restrict these classes due to complexity of the mining problem.

Basic strategy: The most important task of mining algorithm is to generate a candidate set for the target. Because the target class is restricted to tree, it is easy to check if a candidate is embedded in the target. We mention important strategies for candidate generation. *Rightmost expansion* [4, 73] generates subtrees of size $n + 1$ from frequent subtrees of size n with the rightmost branching model: Let T be an ordered tree and P be the rightmost path in T. A new node v is defined as the rightmost child of u for any node u in P. We note that this technique generates all potentially frequent patterns with no duplication thanks to the restriction of order. Figure 9 shows an enumeration tree for (unlabeled) ordered trees by rightmost expansion. In generating unordered tree patterns, the number of ordered trees equivalent to an unordered one of size n is $\Omega(n!)$ in the worst case. Thus, we need to avoid generating such duplicate candidates using the *rightmost expansion with depth sequence*; i.e., unordered trees have canonical trees based on depth-first label sequences [6]. The rightmost expansion based on depth sequences has been widely used [52, 53, 65]. Although there are other strategies as shown in [33], e.g.,

---

[3] For other classes of trees, see e.g., [33].

Fig. 9 Enumeration tree of
the ordered trees by rightmost
expansion

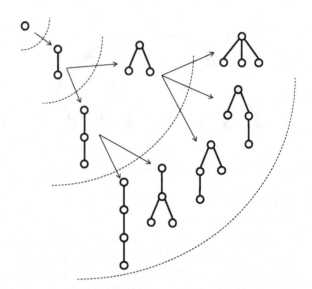

equivalent class-based expansion [74, 75] and right-and-left tree join method [31],
we focus on the above two because of the scalability.

FREQT: We give the algorithm for mining frequent ordered tree patterns [4].
The strategy of this algorithm is based on a simple fact: *any ordered tree of size*
$n + 1$ *is obtained by adding a rightmost edge to a tree of size* n. Fortunately, this
expansion defines an enumeration of all (labeled) ordered trees, called the rightmost expansion; i.e., any tree of size $n + 1$ is generated by a tree of size n. No two
different trees generate a same expansion. Let T be an ordered tree and u be the
node in the rightmost path of T such that the depth of u is d. When adding node v
labeled ℓ to T as the leftmost child of u, the resulting tree T' is called (d, ℓ)
expansion of T. Figure 10 is an outline of FREQT, where the adoptive computation of RMO_k from RMO_{k-1} is derived by the following characteristics. For every
$p \geq 0$ and a node v, the pth parent of v is the ancestor u of v such that the length of
the path from u to v is $p + 1$. Let S be an ordered tree pattern and T be a (d, ℓ)
expansion. For each node u in the tree representing D, $u \in RMO(T)$ iff there exists
$v \in RMO(S)$ such that u labeled by ℓ is a younger sibling of a node that is the
$(d_{max} - d)$th parent of v. Although we omit details of the critical part, Asai
et al. [4, 5] showed that RMO_k is effectively computable from RMO_{k-1} with no
duplication and application to streaming data. The time complexity of FREQT is
bounded by $O(k^2 bLN)$, where k is the maximum size of the frequent patterns, b is
the maximum branching degree of D, L is the number of labels, and N is the sum
of the lengths of the rightmost occurrence lists of frequent patterns. Relative to
FREQT based on the breadth-first approach, Zaki [73] developed a similar algorithm, TreeMiner, based on the depth-first approach. In addition to rightmost
expansion, TreeMiner utilizes a method to prune candidates by *prefix equivalence
class*: let T be a pattern tree of size $n - 1$. All patterns in the set of possible (d, ℓ)

Algorithm FREQT
Input: database D, minimum support σ.
Output: all σ-frequent patterns in D.
1: compute F_1 and RMO_1; set $k = 2$;
2: **while**($F_{k-1} \neq \emptyset$) **do**
    2-1: compute (C_k, RMO_k) from (C_{k-1}, RMO_{k-1}) by
        $C_k = \emptyset$ and $RMO_k = \emptyset$;
        for each($S \in F_{k-1}$ and rightmost expansion T of S) compute $RMO_k(T)$;
        set $C_k = C_k \cup \{T\}$;
    2-2: set $F_k = \emptyset$;
    2-3: **for each**($T \in C_k$) compute $freq_D(T)$ from $RMO_k(T)$;
        if($freq_D(T) \geq \sigma$) set $F_k = F_{k-1} \cup \{T\}$;
        set $k = k + 1$;
3: output $F_1 \cup \cdots \cup F_{k-1}$;

Fig. 10 Outline of FREQT. F_k is the set of σ-frequent patterns of size k in D and the frequency of T in D is denoted by $freq_D(T)$. $RMO_k(T)$ is the set of occurrences of leftmost leaves of pattern T of size k and $RMO_k = \sum_T RMO_k(T)$

expansions of T have the pattern T as the common prefix. TreeMiner reduces the range of occurrences of patterns by the equivalence class.

Extension of FREQT: We summarize two applications of FREQT: *optimal pattern discovery* and *unordered tree mining* [1, 20, 47]. Let D be a set of data, whose elements D_is are called documents (a document is considered a tree in our problem). $S \subseteq D \times \{0, 1\}$ is a set of labeled examples such that exactly one of $(D_i, 1)$ and $(D_i, 0)$ is defined for each $D_i \in D$. D_i is called positive if $(D_i, 1)$ is defined and negative otherwise. Let T be a pattern and a matching function $M(T, D_i) \in \{0, 1\}$ is defined by an appropriate meaning; e.g., an embedding of T into D_i. By (S, T), we can obtain a contingency table (N_1, N_0, M_1, M_0) such that N_1 is the number of positive examples, $N_0 = |D| - N_1$, and M_1 (resp., M_0) are the number of matched (resp., unmatched) positive examples. Then, for an appropriate function ϕ (e.g., the information entropy [47]), the optimization pattern discovery is the problem for minimizing the object function OPD:

$$OPD(S, T, \phi) = \phi(M_1/N_1)N_1 + \phi(M_0/N_0)N_0.$$

Abe et al. extended FREQT to the optimal pattern discovery and proposed an efficient algorithm OPTT [1] to mine optimal patterns from given data D of ordered trees. When we fix the size of maximum patterns and the number of labels, the time complexity of OPTT is $O(b^k N)$, where k is the maximum size of the frequent patterns, b is the maximum branching degree of $D_i \in D$, and $N = |D|$. On the other hand, when k is unbounded, this problem is NP-hard to approximate within a constant factor under the assumption $\text{P} \neq \text{NP}$ [1].

Another application of FREQT was provided for unordered tree mining. Nakano and Uno [49] extended FREQT to unordered tree with $O(1)$ time delay. Using this algorithm, Asai et al. developed an efficient algorithm UNOT [6] for finding unordered frequent patterns. For this enumeration problem, we adopt a canonical form of unordered tree, which is identical to an ordered tree, as a representative. Such canonical form is defined as follows. For a tree T with n-node, a sequence $C(T) = ((d_1, \ell_1), (d_2, \ell_2), \ldots, (d_n, \ell_n))$ is called the depth-label sequence of T, where d_i and ℓ_i is the depth and label of ith visited node by depth-first order. Generally, there exist two or more different $C(T)$s for an unordered tree T. We then consider the ordered tree T' identical to T such that $C(T')$ is lexicographically first. This order is denoted by $\geq_{\text{lex}}$. The ordered tree T' is the canonical form of T; i.e., the set of unordered trees identical to T is named T'. The aim of UNOT is to enumerate canonical forms corresponding to the set of all frequent unordered tree patterns. Nakano and Uno gave the characteristic of the canonical patterns, called the left-heavy condition: An ordered tree T is canonical iff for any $v_1, v_2 \in V(T)$ such that v_1, v_2 are sibling, it holds $C(T(v_1)) \geq_{\text{lex}} C(T(v_2))$, where $T(v)$ denotes the bottom-up subtree with root v. By this, we obtain the following condition. For a canonical T, any T' obtained by removing the rightmost leaf of T is also canonical. This condition defines the enumeration tree of unordered trees similarly to the ordered case. However, the useful monotonicity for expanding the enumeration tree is not preserved in unordered cases, because a rightmost expansion of canonical form is not exclusively canonical. Nakano and Uno overcame this difficulty by local search around a special node in the current rightmost path, and their algorithm outputs all canonical forms in $O(1)$ delay time per tree with no duplication. In addition, some readers might be surprised by two facts: the time delay for reporting any canonical tree T is $O(1)$ and the size of T is not bounded by $O(1)$. The trick is to design a nice enumeration understanding that the size of the difference of T and the next T' is bounded by $O(1)$.

3.2 Cohesive Subgraphs

We next consider graph decomposition by dense subgraphs. k-core, n-clique, n-clans, n-cubes, k-plexes, and other notions were introduced to identify cohesive subgroups within social networks [72]. Only the problem of finding k-cores is efficiently solvable, because others require at least quadratic time in input size. Thus, we begin with graph decomposition based on k-cores toward other related and tractable decompositions.

k-core: We passed over the notion of k-cores in previous sections; let us now recall the definition. For a graph H, an induced subgraph $G = (V, E)$ of H is a *k-core* if for any $v \in V$, $d_G(v) \geq k$ and V is maximum with this property. Figure 11 shows an example of k-core decomposition of a whole graph with up to $k = 3$. For this notion, we should note that the classes of k-cores are nested, a core is not necessarily connected, and a core is not necessarily dense subgraphs. The k-core decomposition algorithm [60] is widely used to partition large graphs into a hierarchical layer thanks to its efficiency. The original algorithm is shown in Fig. 12. Actually, as shown in [8], by ordering all nodes in increasing order of

Fig. 11 1-, 2-, and 3-cores

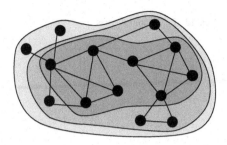

Fig. 12 k-core algorithm

Algorithm for finding k-cores
Input: a graph G.
Output: k-th core.
$k = 0$
$G_0 = G$
while$(|G_k| > 0)$ **do**
    while $\exists v \in V(G_k)$ such that $d(v) \leq k$ **do**
        delete v from G_k
    $G_{k+1} = G_k$
    $k = k + 1$

their degrees, the improved algorithm runs in $O(m)$ time, where m is the number of edges for the given graph.

DN-graph: We next summarize the *dense neighborhood graph* (DN-graph) introduced by Wang et al. [70]. Intuitively, a dense pattern in a graph contains a set of highly relevant nodes, and they often share many common neighbors. The *neighborhood* of node v is the set of nodes adjacent to v, denoted by $N(v)$. If two nodes share some common neighbors, the set of such common nodes is called the *joint* neighborhood of the pair of nodes. For a subgraph G' of G, the neighborhood of G' is the set of nodes in $V(G) - V(G')$ adjacent to a node in $V(G')$. Inside a graph, the measurement of minimal joint neighborhood size between any adjacent node pair is denoted by λ. Using this parameter λ, DN-graph is defined. A DN-graph with λ is a connected subgraph $G'(V', E')$ of a graph $G = (V, E)$ that satisfies: (1) For any $\{u, v\} \in E'$, there exist at least λ nodes adjacent to both u and v. (2) For any $v \in V - V'$, $\lambda(V' \cup \{v\}) < \lambda$ and; For any $v \in V'$, $\lambda(V' - \{v\}) \leq \lambda$, where $\lambda(V' \cup \{v\})$ denotes the measurement for the subgraph induced by the set $V' \cup \{v\}$, and $\lambda(V' - \{v\})$ is similar. Because the optimization problem for DN-graph is NP-hard, Wang et al. [70] proposed several algorithms approximating the optimum solution based on graph triangulation [59] for listing all triangles in $O(m^{1.5})$ time and $O(m + n)$ space. On off-line setting, the time complexity of their algorithm is bounded by $O(km^{1.5} \log n)$ and the space complexity is $O(n \log n + m)$.

Fig. 13 2-, 3-, and 4-truss

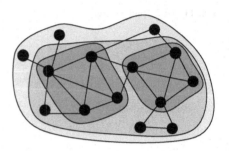

k-truss: Recently, graph decomposition based on another type of cohesive subgraphs, called k-truss, has been addressed by researchers [18, 69]. Given a graph G, the k-truss of G is the largest subgraph of G in which any edge is contained in at least $k - 2$ triangles within the subgraph. The problem of truss decomposition of G is to find k-trusses in G for all k. It is trivial that 2-truss is G itself. Figure 13 shows an example of k-truss decomposition of whole graph with up to $k = 3$. This notion is similar to k-core, which is the largest subgraph in which any node has a degree of at least k within the subgraph; however, as indicated in [18], these notions are incomparable, because, e.g., a k-truss is a $(k - 1)$-core and not vice-versa. We outline the truss decomposition. Given a graph $G = (V, E)$, the *support of an edge* $e = \{u, v\} \in E$, denoted by $support(e, G)$, is defined as

$$|\{\triangle_{uvw} \mid \triangle_{uvw} \in \triangle_G\}|,$$

where $\triangle_{uvw}$ is identical to K_3 consisting of nodes u, v, w, and $\triangle_G$ is the set of triangles in G. The support of an edge $e \in E$ is the number of triangles in G containing e. Then, truss decomposition is to find the k-truss of G, denoted by T_k, which is the largest subgraph of G such that $support(e) \geq k - 2$ for any $e \in E(T_k)$ for all k. For the decomposition problem, Cohen [18] developed the first polynomial time algorithm shown in Fig. 14. This algorithm requires $O(m + n)$ space to keep the input graph as well as the support of all edges in memory and $O(\sum_{v \in V} d(v)^2)$ time, which is quite expensive for dense graphs. Wang et al. [69] improved this algorithm for performing lightweight truss decomposition. Using the in-memory triangle counting algorithm [37], the improved algorithm achieved $O(m^{1.5})$ time complexity with the same space to that of Cohen.

3.3 Cliques

Clique is the most important community, because its density is maximum among graphs of the same size. However, detecting clique is harder than with previous pattern extraction. We next focus on detecting maximal cliques by state-of-the-art algorithms.

Algorithm for truss decomposition
Input: $G = (V, E)$.
Output: k-truss for $3 \le k \le k_{max}$.
$k \leftarrow 3$
for each $e = \{u, v\} \in E$ **do**
$\quad support(e) = |N(u) \cap N(v)|$
while($\exists e = \{u, v\}$ such that $support(e) < k - 2$) **do**
$\quad W \leftarrow N(u) \cap N(v)$
$\quad$ **for** each $e' = \{u, w\}$ or $e' = \{v, w\}$, where $w \in W$, **do**
$\quad\quad support(e') \leftarrow sup(e') - 1$
$\quad$ remove e from G
output G as the k-truss
if($E \ne \emptyset$)
$\quad k \leftarrow k + 1$
$\quad$ **goto while**

Fig. 14 k-truss algorithm

Related work: A subset $K \subseteq V$ of $V(G)$ is called a clique in a graph G if any two nodes in K are adjacent. A clique is called *maximal* if it is not contained in any other clique. A clique in a bipartite graph is called a bipartite clique. We consider the problem for enumerating all maximal cliques (in polynomial time delay if possible). Tsukiyama et al. [67] proposed the $O(nm)$ time delay algorithm for maximal *independent sets of G*, where $n = |V(G)|$ and $m = |E(G)|$. Using the straightforward relation between independent sets and cliques, this algorithm achieved the same time complexity for maximal clique enumeration. This algorithm was improved by Chiba and Nishizeki [16]. Makino and Uno [43] developed new algorithms for general graphs in $O(\Delta^4)$ time delay and for bipartite graphs in $O(\Delta^3)$ time delay with $O(nm)$ preprocessing time, $O(n + m)$ space complexity, and $\Delta = \Delta(G)$. These algorithms are especially efficient for sparse graphs. After their work, Tomita et al. [66] proposed an optimal algorithm for maximal clique enumeration whose total running time is $O(3^{n/3})$. We note that this is the best one could hope for with n, because there exists up to $\mu = 3^{n/3}$ cliques in a graph of size n [46]. In contrast with Tomita et al.'s algorithm, that of Tsukiyama et al. requires $O(nm\mu)$ total time and that of Makino and Uno requires $\Delta^4\mu$ total time. Moreover, the last two algorithms were implemented and proved to be efficient for large data. From these algorithms we focus on the bipartite clique enumeration and summarize the basic idea because the bipartite clique is closely related to data mining from transaction data, which is the next topic.

Basic technique: We summarize the technique by Tsukiyama et al. [67] for enumerating all maximal cliques. Many algorithms depend on this result. Let $G = (V, E)$ be a graph such that $V = \{v_1, \dots, v_n\}$ and $|E| = m$. For $v \in V$, let $A(v) = \{u \in V \mid \{u, v\} \in E\}$, i.e., the set of nodes adjacent to v. For a subset $S \subseteq V$ and i, let $S_{\le i} = S \cap \{v_1, \dots, v_i\}$. For two sets of nodes, X and Y, we write $X \ge_{lex} Y$ (X is lexicographically larger than Y) if the node with the smallest index in $X \oplus Y$ is in X. For a clique K of G, let $C(K)$ denote the maximal clique such that $C(K) \geq_{\text{lex}} K'$ for any maximal clique K' satisfying $K \subseteq K'$. We note that there is the monotonic relation such that $K \geq_{\text{lex}} C(K)$ never holds. Using this binary relation, we can define a tree structure corresponding an enumeration of all maximal cliques with no duplication. Let $K_0 \geq_{\text{lex}} K$ for any maximal clique K. For a K ($\neq K_0$), define the parent of K by $P(K) = C(K_{\leq i-1})$ such that i is the maximum index satisfying $C(K_{\leq i-1}) \neq K$. The index i is called *parent index* of K, denoted by $i(K)$. Since $P(K) \geq_{\text{lex}} K$, this binary relation on maximal cliques represents an enumeration tree for maximal cliques in G rooted by K_0. To traverse this tree, it is easy to compute $P(K)$ from a given K by checking all nodes adjacent to a node in K. To obtain a child K' of a maximal clique K in this tree, we need the following characteristics proved by Tsukiyama et al. [67] and modified by Makino and Uno [43]: For the set

$$K[i] = C((K_{\leq i} \cap A(v_i)) \cup \{v_i\}),$$

$K[i]$ is a child of K iff

(1) $v_i \notin K$,
(2) $i > i(K)$,
(3) $K[i]_{\leq i-1} = K_{\leq i} \cap A(v_i)$, and
(4) $K_{\leq i} = C(K_{\leq i} \cap A(v_i))_{\leq i}$.

The index i satisfying the above conditions is the parent index of $K[i]$. With this result, we can compute all children of K in $O(nm)$ time delay in $O(n+m)$ space. On the other hand, we often find important structures in data as binary relations, e.g., Kleinberg's *hubs and authorities* [35], represented by bipartite graph $G = (V_1 \cup V_2, E)$. G is restricted so that there is no edge connecting $u \in V_1$ and $v \in V_2$, but this condition does not make the problem easy, because any graph $G_1 = (V_1, E_1)$ is transformed to a bipartite graph $G_2 = (V_1 \cup V_2, E_2)$, where V_1 is a copy of V_2 and $e = \{u, v\} \in E_2$ iff $e \in E_1$. Thus, the problem of enumerating bipartite cliques is sufficiently general. For a maximal bipartite clique enumeration, Makino and Uno proposed an algorithm to enumerate maximal bipartite cliques in $O(\Delta^3)$ time delay with $O(nm)$ preprocessing time (see [43] for detail).

Applications: There is a one-to-one correspondence between maximal bipartite cliques in a bipartite graph and closed itemsets in transaction data [76]. As we have seen in the above, maximal cliques are efficiently enumerable. We show that this algorithm can be applicable to an important data mining problem. Let I be a set of *items* and $T \subset 2^I$, where $\tau \in T$ is called a *transaction*. For a constant $\alpha \geq 0$, $S \subset I$ is called α-frequent if at least α transactions of T include S. Finding all α-frequent sets is one of most important problems in data mining, because it characterizes T by association rules [13]. It is easy to enumerate all frequent sets using the monotonicity on frequency. However, there are too many frequent sets to enumerate. To avoid the explosion of outputs, it is useful to restrict patterns to be maximal frequent. Unfortunately, this enumeration problem is known as a

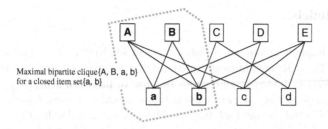

Transactions A, B, C, D, E on items $\{a, b, c, d\}$:	All maximal bipartite cliques with at least 2 transactions and 2 items:
A = {a, b, c}	**{A, B, a, b}**
B = {a, b}	{A, D, a, c}
C = {b, d}	{A, E, b, c}
D = {a, c}	{B, D, a, b}
E = {b, c, d}	{C, E, b, d}

Fig. 15 Relation between maximal bipartite clique and closed itemset. In this example, a transaction is a subset of $I = \{a, b, c, d\}$. There are five transactions A, B, C, D, and E. An item set $P \subseteq I$ is called frequent if the number of transactions including P is larger than a threshold, and is called closed if there is no super set of P included by the same transactions. A closed item set is corresponding to a maximal bipartite clique. The bipartite graph $G = (V_1 \cup V_2, E)$ is figured for $V_1 = \{A, B, C, D, E\}$, $V_2 = \{a, b, c, d\}$, and $\{u, v\} \in E$ iff $u \in V_1$, $v \in V_2$, and the transaction u contains the item v

computationally hard problem. Instead of this problem, many researchers have studied the enumeration of *closed* frequent sets S; there is no other superset S' of S such that $S' \subseteq \tau$ for any $\tau \in \mathcal{T}$. We note that the number of closed itemsets is usually much smaller than that of frequent sets. We then show a one-to-one correspondence between maximal bipartite cliques and closed itemsets. Together with the enumeration algorithm for maximal bipartite cliques, the closed itemsets enumeration problem is efficiently solvable. For an I and $\mathcal{T} \subset 2^I$, we can construct a bipartite graph $G = (V_1 \cup V_2, E)$ whose maximal bipartite graphs correspond to closed itemsets in $\mathcal{T}$ in one-to-one:

- V_1 is the set of u corresponding to a member of $\mathcal{T}$,
- V_2 is the set of v corresponding to a member of I, and
- $e = \{u, v\} \in E$ iff the item corresponding to v is included in the transaction corresponding to u.

This correspondence can be easily understood by the illustration in Fig. 15. Uno et al. [68] proposed a first polynomial time delay algorithm LCM, directly enumerating frequent closed itemsets, whereas Makino and Uno's algorithm enumerates maximal bipartite cliques. For the detail of association rule mining and its applications, see the survey [13].

4 Implicit Models

A definition of implicit communities was proposed by Kumar et al. [36] as a community is a dense directed bipartite subgraph that contains a clique of certain size. This definition is, however, ambiguous because of the word *dense*. For this problem, we mainly focus on two unambiguous models: the modularity by Newman et al. [51] and the network flow by Flake et al. [23]. Modularity is an index to measure the distance between extracted community and random graph of same size. Unfortunately, finding a best solution for this problem is computationally hard [11]. Thus, we usually use an approximation of ideal modularity [10, 17, 50]. On the other hand, the network flow model is based on the maximum flow problem, which is one of the tractable problems. Using them, we obtained unambiguous definitions for implicit communities.

4.1 Modularity and Its Approximation

Top-down clustering: The definition of modularity Q for a set of communities C was given in Sect. 2.3. Since maximizing Q corresponding to optimal communities is NP-hard [11], many approximation algorithms have been proposed. Newman and Girvan [51] proposed a top-down clustering algorithm by *betweenness* (Sect. 2.3) together with Q, where betweenness means *edge betweenness*. Originally, the node betweenness [25] is the fraction of shortest paths containing node v defined by

$$Bet(v) = \sum_{s,t \in V(G)} \frac{SP_v(s,t)}{SP(s,t)},$$

where $SP(s,t)$ is the number of shortest paths from node s to t, and $SP_v(s,t)$ is the number of those paths containing node v. Newman and Girvan [51] extended this definition to the case of edges, $Bet(e)$, defining the edge betweenness of edge e as the number of shortest paths including the edges in question. The clustering algorithm based on betweenness is summarized below, and obtained clusters, i.e., a set of communities, are evaluated by modularity Q.

1. Compute $Bet(e)$ for any edge $e \in V(E)$.
2. Remove the edge $e \in E$ with the highest $Bet(e)$.
3. Compute the betweenness of all edges for the resulting graph.
4. Repeat the above process until no edge remains.

Bottom-up clustering: Newman and Girvan's algorithm requires $O(nm^2)$ time with $n = |V(G)|$ and $m = |E(G)|$, which is inefficient for large-scale graphs. To address this difficulty, Newman proposed a bottom-up algorithm using $\Delta(i,j)$, defined as

$$\Delta(i,j) = 2(e(i,j) - d_i d_j),$$

where $e(i,j)$ is the number of edges between communities i,j, and d_i is the sum of $d(v)$ for all node v in community i. Intuitively, $\Delta(i,j)$ is the variation of Q when communities i and j are merged. By this measure, Newman's algorithm is summarized below.

1. Initially, every node is a cluster itself with $Q = 0$.
2. Compute $\Delta(i,j)$ for any pair (i,j).
3. Make cluster k merging i and j for the highest $\Delta(i,j)$.
4. Compute Q for the new cluster k.
5. Repeat 2–4 while there are two clusters or more.
6. Output a cluster k with maximum Q.

Newman's algorithm requires $O((n + m)n)$, which is faster than Newman and Girvan's algorithm in $O(nm^2)$ time. We note that $m = \Omega(n^2)$ for dense graphs, and $m = O(n)$ for sparse graphs. For sparse graphs, the time complexity of Newman's algorithm is $O(n^2)$; this was improved to $O(n \log^2 n)$ by Clauset et al. [17] using heap structures. For this approach using modularity, other practical algorithms were proposed (see e.g., [10]).

4.2 Network Flow

According to Flake et al. [23], a community is *a set of nodes in which each member has at least as many edges connecting to members as it does to non-members*. This definition is clear and in any case, we can determine whether or not a candidate is a community. Generally, the problem of finding such a subset is computationally hard; however, exploiting various properties of graph structures allows us to reduce the problem of extracting communities to the tractable max-flow problem [29]. They showed an interesting relation between such dense subgraphs and a set of saturated edges in maximum flow, which is efficiently computable. Such communities extracted by the max-flow algorithm are called max-flow communities. This method is widely applied to actual web graphs [24] and other domains; e.g., see [48].

As we have seen, a community is a set of nodes, $C \subseteq V$, such that for all nodes $v \in V$, v has at least as many edges connecting to nodes in C as it does to node in $V - C$. For this optimization problem, which is computationally hard, Flake et al. [23] showed that communities can be approximately extracted as min. cuts as follows. Let $\#s$ refer to the number of edges between s and all nodes in $C - \{s\}$, and let $\#t$ refer to the number of edges between t and $(V - C - t)$. A community C can be identified by calculating the $s - t$ min. cut of G with s and t being used as the source and sink respectively, provided that both $\#s$ and $\#t$ exceed the cut set

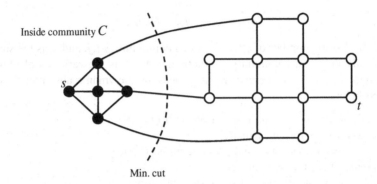

Fig. 16 Community extracted by minimum cut

size. After the cut, nodes that are reachable from s are in the community C. This is illustrated in Fig. 16.

Next we recall the *network flow* defined on digraphs, which is identical to the cut problem. Given a digraph $G = (V, E)$ with edge capacity $c(e) \in Z^+$ and two nodes $s, t \in V$ called *source* and *sink* respectively, a flow in G is a function $f : E \to R^+$ satisfying the capacity and conservation constraints:

(capacity constraint) $f(e) \le c(e)$ for any $e \in E$

(conservation constraint) $\displaystyle\sum_{e \in IN(v)} f(e) = \sum_{e \in OUT(v)} f(e)$ for any $v \in V - \{s, t\}$

where $IN(v) = \{e \in E \mid end(e) = v\}$ and $OUT(v) = \{e \in E \mid start(e) = v\}$; i.e., $IN(v)$ is the set of edges directed to v and $OUT(v)$ is analogous. Intuitively, the maximum flow problem asks you how much water we can move from the source to the sink. The Ford and Fulkerson's max. flow-min. cut theorem (see e.g., [19]) shows that the max. flow of a network is identical to a min. cut that partitions $V = V_s \cup V_t$ such that $s \in V_s$ and $t \in V_t$.

A web graph is considered a digraph. However, we cannot assume $s - t$ network representing the graph, since any node must be reachable from s and t must be reachable from any node. Beyond that, we cannot handle the whole web graph due to its size. For this problem, we consider a partial web graph crawled from seed web sites selected beforehand. This strategy is shown in Fig. 17. Using the max. flow-based web crawler, the extraction algorithm approximates a community by directing a focused web crawler along hyper-link paths that are regraded to be highly relevant. Flake et al. [23] reported that discovered communities are highly dense in the sense that members of the community are more tightly coupled to each other than to non-members.

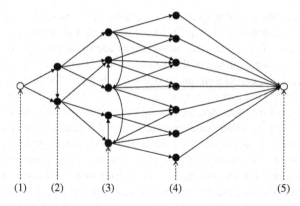

Fig. 17 Focused partial web graph for community extraction: (*1*) the virtual source node directed to any seed node; (*2*) seed nodes selected beforehand; (*3*) nodes directed from a seed; (*4*) other nodes directed from a node of (*3*) not included in (*2*) and (*3*); and (*5*) the virtual sink node directed from any node of (*4*). For each edge, a same unit capacity is defined for max. flow algorithm

5 Beyond Static Patterns

The community extraction algorithms discussed in Sects. 3 and 4 are assumed to have just one static graph as their inputs. In the real world, social relationships keep changing over time. A community structure can also change; i.e., shrink, split, grow, and merge. To understand the behaviors of social networks over time, we therefore need methods to trace changes in a series of graphs.

5.1 Sequential Pattern Mining in Data Stream

Framework: We have seen the relation between mining in graphs and transactions; e.g., a maximal clique in bipartite graph corresponds to a closed itemsets in the transaction database. In this section we extend this framework to (multiple) data streams to handle pattern discovery in web-log, micro-blog, and other time-series data.

A framework of sequential mining was introduced by Agrawal and Srikant [2, 63]. Let D be a set of transactions such that each $T \in D$ consists of id, timestamp, and an itemset, i.e., a set of items constructing the domain of transaction. Let $I = \{i_1, \ldots, i_m\}$ be a set of items. An itemset is a non-empty subset of I. A sequence $S = (s_1, \ldots, s_n)$ is a set of items ordered by their timestamps, where each s_i is an itemset. A k-sequence is a sequence of k itemsets. A sequence $S' = (s'_1, \ldots, s'_m)$ is a subsequence of S, denoted by $S' \prec S$, if there exist integers $i_1 < \cdots < i_m$ such that $s'_1 \subseteq s_{i_1}, \ldots, s'_m \subseteq s_{i_m}$, i.e., S' is embedded into S. All transactions with a same id are grouped together and sorted in increasing order of id. A sequence of the sorted transactions is called *data sequence*. Let $occ(S)$ denote the number of occurrences of S in D, where any S in a data sequence is counted only once for $occ(S)$ even if other occurrences are discovered. A data sequence contains S if S is a subsequence of the data sequence. To decide whether or not S is frequent, a threshold, called a *minimum support* σ, is given for the user beforehand, and S is said to be *frequent* if $occ(S) \geq \sigma$.

On the standard frequent itemset mining covered in previous sections, patterns are monotonic. On the other hand, we know that the anti-monotonicity holds for sequential patterns with respect to a constraint [56]. Let us briefly mention the (anti-)monotonicity. A constraint C is *anti-monotonic* if a sequence S satisfying C implies that any non-empty subsequence of S also satisfies C, and C is *monotonic* if a subsequence of S satisfies C implies that any super sequence of S also satisfies C. Given a database, the problem of mining sequential patterns is to find all the frequent sequences with a minimum support σ specified beforehand.

We next set the sequential mining in stream data [58]. Let data stream $DS = B_1, B_2, \ldots, B_n, \ldots$ be an infinite sequence of batches, where each batch B_i is associated with a time period $[a_i, b_i]$ with $a_i < b_i$, and let B_n be the most recent batch for the largest n. Each batch B_k consists of a set of data sequences defined above, i.e., $B_k = \{S_1, S_2, \ldots, S_m\}$. For each data sequence S in B_k, we are provided with its list of itemsets. Of course, we assume that batches do not necessarily have a same size. The size of the data stream at n is defined as $\sum_{1 \leq i \leq n} |B_i|$ for the cardinality $|B_i|$ of B_i.

We define the support of a sequential pattern in stream data as follows: the support of a sequence S at time i is the ratio of the number of data sequences having S at the current time window to the total number of data sequences. Thus, the problem of sequential patterns in stream data is to find all frequent patterns S_k over an arbitrary period $[a_i, b_i]$ qualified by

$$\sum_{t=a_i}^{b_i} occ_t(S_k) \geq \sigma |B_i|.$$

Related work: In the last decade, the main problem of stream data mining is to maintain frequent itemsets over the entire history. Manku and Motowani [44] proposed a first single-pass algorithm for stream data based on the anti-monotonicity. This algorithm computes the support from the start of the stream. Anti-monotonicity (and monotonicity) is included in the constraint model. For an extensive survey of constraints pattern mining; see e.g., [56].

Li et al. [41] proposed another single-pass mining algorithm, which mines the set of all maximal frequent itemsets in landmark windows over streams. A new summary data structure called summary frequent itemset forest was developed to incrementally maintain the essential information about maximal frequent itemsets embedded in the stream so far. This algorithm was further extended [40].

Chi et al. [15] proposed the *closed enumeration tree* to maintain dynamically selected set of itemsets for closed frequent itemsets mining. Raissi et al. [58]

introduced a new framework of sequential pattern mining in stream data and proposed a data structure to dynamically maintain frequent sequential patterns with a fast pruning strategy. At any time, users can issue requests for frequent maximal sequences over an arbitrary time interval.

Snowsill et al. [62] addressed the problem of detecting surprising patterns in large textual data streams using n-gram model and generalized suffix tree annotated with a limited amount of statistical information. We can find many other studies along this line (see e.g. [3, 27, 34, 38]).

Many researchers focused on sequential mining from *multiple data streams*. Oates and Cohen [54] proposed an algorithm to find the dependency rules among stream data in a multiple data streams. Chen et al. [14] proposed an algorithm based on the prefix-projection concept of PrefixSpan [55], handling multiple data streams in a time period and mining sequential patterns in time sequence data stream environment. For other recent results, see e.g., [22].

5.2 Explicit Approaches for Tracing Communities

There are a number of graph-theoretic approaches to identifying community flows over time. In most algorithms, a set of communities is assumed to be given in each snapshot graph using a community extraction algorithm. Therefore, the problem of identifying the community flows is independent from a specific community extraction algorithms, and reduced to the problem of tracing community transitions or mutation over time from the sequence of node sets.

Identifying Communities with Graph Coloring: Berger-Wolf and Saia proposed a framework to analyze dynamic social networks [9]. Following their problem setting, Tantipathananandh et al. formulated the problem of tracing community transitions as a vertex coloring problem [64] in a graph. In their approach, the communities at each time stamp are assumed to be given as disjoint subsets of the node set in each snapshot graph by a certain community detection algorithm, and each node is uniquely labeled, or each node is uniquely identified over time.

Let $X = \{x_1,\ldots,x_n\}$ be a set of individuals occurring in time-series social networks. A subset $C \subseteq X$ is a community of individuals. We assume that a partition P_t of a graph observed at time t is given, where P_t is a set of communities, and any two communities C and C' in P_t are disjoint; i.e., $C \cap C' = \emptyset$ for any $C, C' \in P_t$ if $C \neq C'$. Then, a sequence of partitions observed for time $t = 1,\ldots,T$ is denoted by $H = \langle P_1,\ldots,P_T \rangle$.

In the method, they first transform the sequence of partitions H into a graph $G = (V, E)$. This graph G has an individual vertex $v_{x,t} \in V$ for each individual $x \in X$ and time t, and a group vertex $v_{C,t}$ for each group $C \in P_t$. These vertices are connected with edge $\{v_{x,t}, v_{x,t+1}\} \in E$ for each individual $x \in V$ and time $t \in \{1,\ldots,T-1\}$, and $\{v_{x,t}, v_{C,t}\} \in E$ for each individual $x \in g$ in each partition P_t.

On this graph, they define a *community interpretation* as a vertex coloring of G so that any two communities C and C' does not share the same color at each time. In addition, three criteria are introduced to measure the quality of a community interpretation.

The community tracing problem is defined as a vertex coloring optimization problem. Since the vertex coloring is NP-complete, they introduce efficient greedy heuristics and apply them to relatively small networks.

Identifying Communities with Bipartite Mapping: Greene et al. formulate the problem of tracing communities over time as a weighted bipartite mapping problem [30]. Their model allow communities to be disjoint and overlapped subsets of nodes in snapshot graphs. The main idea of this approach lies in finding a correspondence between two consecutive snapshots of communities $\{C_{1,t}, \ldots, C_{k,t}\}$ and $\{C_{1,t+1}, \ldots, C_{k',t+1}\}$.

They employ a relaxed version of bipartite matching (we refer to it as *bipartite mapping*), which allows many-to-many node correspondences between two sets of nodes in order to identify dynamic events such as community merging and splitting.

A dynamic graph is defined as a sequence of snapshot graphs $\langle G_1, \ldots, G_T \rangle$. For each snapshot graph G_i, communities are extracted using a static community extraction algorithm, and let $C_t = \{C_{t,1}, \ldots, C_{t,k_t}\}$ be the set of communities in each time $t \in \{1, \ldots, T\}$.

A *dynamic community* D_i is defined as a sequence of communities, which represents a track of the same communities over time, and D_i can include at most one community in each time t. The dynamic communities are initialized by $D_i = \langle C_{1,i} \rangle$ for each $C \in C_1 = \{C_{1,1}, \ldots, C_{1,k_1}\}$. A *front* of a dynamic community D_i is the last element of D_i, and denoted by F_i. Figure 18 shows an example of dynamic communities.

$$D_1 = \{C_{11}, C_{21}, C_{31}\} \qquad D_2 = \{C_{12}, C_{21}, C_{31}\}$$
$$D_3 = \{C_{13}, C_{22}, C_{32}\} \qquad D_4 = \{C_{13}, C_{23}\}$$
$$F = \{C_{31}, C_{32}, C_{23}\}$$

The algorithm iteratively attempts to find a weighted bipartite mapping between a set of fronts $F = \{F_1, \ldots, F_{k'}\}$ in dynamic communities and a set of communities

Fig. 18 Four dynamic communities over three time stamps

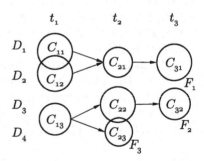

$C_t = \{C_{t,1}, \ldots, C_{t,k_t}\}$ in a snapshot graph G_t with a threshold of edge weight θ. The edge weights are given by the Jaccard coefficient as follows:

$$\text{sim } (C_{t,j}, F_i) = \frac{|C_{t,j} \cap F_i|}{|C_{t,j} \cup F_i|}.$$

A new dynamic community is created when there is no mapping for $C_{t,j}$ or one existing dynamic community is mapped to two or more communities in C_t (i.e, split of a dynamic community). They show that this algorithm runs in almost linear time to the number of nodes.

5.3 Implicit Approaches for Tracing Communities

In the explicit models, snapshot communities over time are assumed to be given, and the main task is to identify the relation among communities in different time stamps mainly from the graph/set theoretic points of view. In contrast to the explicit models, this section shows several studies on implicit models of dynamic graphs. These approaches often incorporate a community extraction method to identify the transitions of graphs with a probabilistic process or a learning algorithm.

PageRank-based Tree Learning Method: Qui and Lin studied on an evolution of hierarchical structures of dynamic social networks in [57]. Their approach combines the hierarchical community extraction using a PageRank [12]-based algorithm and a tree learning algorithm using tree edit distance.

First, each snapshot of the social network in each time stamp is transformed into a static hierarchical structure (referred to as community tree) so that a node with a higher PageRank-based score comes to be an upper node of the other nodes with lower scores. Next, two consecutive community trees over time are reconstructed as an evolving community tree so that it minimizes the tree edit distance against the current and previous community trees. The method is applied to the Enron dataset and shown to reveal a real organization well.

Probabilistic Models: Many recent studies on dynamic networks have been made based on probabilistic models. Most of these methods unify the tasks of extracting communities and tracing their evolution over time.

Lin et al. introduced a framework for analyzing communities and their evolution in dynamic networks [42] using non-negative matrix factorization (NNMF). Their method employs soft clustering on the time series networks in which each node has the probability of belonging to clusters.

Fu et al. proposed a very general probabilistic model for analyzing dynamic networks with a dynamic mixed membership stochastic block model based on a latent space model. This model assumes the functional roles of network actors transiting over time [26].

Ishiguro et al. focuses on modeling drastic change in dynamic networks caused by splitting or merging events but most existing probabilistic methods model gradual changes of network structures [32]. Their approach employs the infinite HMM model to infer the transitions of communities without specifying the their number.

6 Conclusion

Pattern extraction from graphs has been widely applied to the field of social network analysis. From the modeling points of view, there are mainly two approaches: explicit modeling and implicit modeling. We emphasize explicit modeling rather than implicit modeling because it has yet to be developed as compared with implicit modeling based on highly-developed statistical machine learning methods. The advantage of explicit modeling is, first, that the algorithms are, in general, much simpler; and, second, that the interpretation of the results is easier, that is, we can easily understand why the results are obtained thanks to the explicit models.

References

1. Abe, K., Kawasoe, S., Asai, T., Arimura, H., Arikawa, S.: Optimized substructure discovery for semi-structured data. In: PKDD, pp. 1–14 (2002)
2. Agrawal, R., Srikant, R.: Mining sequential patterns. In: ICDE, pp. 3–14 (1995)
3. Allan, J., Papka, R., Lavrenko, V.: On-line new event detection and tracking. In: SIGIR, pp. 37–45 (1998)
4. Asai, T., Abe, K., Kawasoe, S., Arimura, H., Sakamoto, H., Arikawa., S.: Efficient substructure discovery from large semi-structured data. In: SDM, pp. 158–174 (2002)
5. Asai, T., Arimura, H., Abe, K., Kawasoe, S., Arikawa., S.: Online algorithms for mining semi-structured data stream. In: ICDM, pp. 27–34 (2002)
6. Asai, T., Arimura, H., Uno, T., Nakano, S.: Discovering frequent substructures in large unordered trees. In: Discovery Science, pp. 47–61 (2003)
7. Backstrom, L., Huttenlocher, D.P., Kleinberg, J.M., Lan, X.: Group formation in large social networks: membership, growth, and evolution. In: KDD, pp. 44–54 (2006)
8. Batagelj, V., Zaversnik, M.: An O(m) algorithm for cores decomposition of networks. arXiv, preprint cs/0310049 (2003)
9. Berger-Wolf, T.Y., Saia, J.: A framework for analysis of dynamic social networks. In: Proceedings of the 12th ACM SIGKDD international conference on knowledge discovery and data mining, pp. 523–528. ACM (2006)
10. Blondel, V.D., Guillaume, J.-L., Lambiotte, R., Lefebvre, E.: Fast unfolding of communities in large networks. J. Stat. Mech. Theory Exp. 2008, 10008 (2008)
11. Brandes, U., Delling, D., Gaertler, M., Gorke, R., Hoefer, M., Nikoloski, Z., Wagner, D.: On modularity clustering. IEEE Trans. Knowl. Data Eng. 20, 172–188 (2008)
12. Brin, S., Page, L.: The anatomy of a large-scale hypertextual web search engine. Comput. Networks ISDN Syst. 30(1), 107–117 (1998)
13. Ceglar, A., Roddick, J.F.: Association mining. ACM Comput. Surv. 38(2), 5 (2006)

14. Chen, G., Wu, X., Zhu, X.: Sequential pattern mining in multiple streams. In: ICDM, pp. 27–30 (2005)
15. Chi, Y., Wang, H., Yu, P.S., Muntz, R.R.: Moment: maintaining closed frequent itemsets over a stream sliding window. In: ICDM, pp. 59–66 (2004)
16. Chiba, N. Nishizeki, T.: Arboricity and subgraph listing algorithms. SIAM J. Comput. 14(1) 210–223 (1985)
17. Clauset, A., Newman, M.E.J., Moore, C.: Finding community structure in very large networks. Phys. Rev. E **70**, 66111–66117 (2004)
18. Cohen, J.D.: Trusses: cohesive subgraphs for social network analysis. National Security Agency Technical Report (2008)
19. Cormen, T.H., Leiserson, C.E., Rivest, R.L.: Introduction to Algorithms. MIT press and McGraw-Hill Book Co., Cambridge (1992)
20. Dehaspe, L., Toivonen, H., King, R.D.: Finding frequent substructures in chemical compounds. In: KDD, pp. 30–36 (1998)
21. Diestel, R.: Graph Theory. Springer, Heidelberg (2000)
22. Ezeife, C.I., Monwar, M.: SSM: a frequent sequential data stream patterns miner. In: CIDM, pp. 120–126 (2007)
23. Flake, G.W., Lawrence, S., Giles, C.L.: Efficient identification of web communities. In: KDD, pp. 150–160 (2000)
24. Flake, G.W., Lawrence, S., Giles, C.L., Coetzee, F.: Self-organization and idenfitication of web communities. IEEE Comput. **35**(3), 66–71 (2002)
25. Freeman, L.C.: A set of measures of cenrrality based upon betweenness. Sociometry **40**, 35–41 (1977)
26. Fu, W., Song, L., Xing, E.P.: Dynamic mixed membership blockmodel for evolving networks. In: Proceedings of the 26th annual international conference on machine learning, pp. 329–336. ACM (2009)
27. Fung, G.P.C., Yu, J.X., Yu, P.S., Lu, H.: Parameter free bursty events detection in text streams. In: VLDB, pp. 181–192 (2005)
28. Girvan, M., Newman, M.E.J.: Community structure in social and biological networks. PNAS **99**(12), 7821–7826 (2002)
29. Goldberg, A.V., Tarjan, R.E.: A new approach to the maximal flow problem. In: STOC, pp. 136–146 (1986)
30. Greene, D., Doyle, D., Cunningham, P.: Tracking the evolution of communities in dynamic social networks. In: 2010 international conference on advances in social networks analysis and mining (ASONAM), pp. 176–183. IEEE (2010)
31. Hido, S., Kawano, H.: AMIOT: Induced ordered tree mining in tree-structured databases. In: ICDM, pp. 170–177 (2005)
32. Ishiguro, K., Iwata, T., Ueda, N., Tenenbaum, J.: Dynamic infinite relational model for time-varying relational data analysis. Adv. Neural Inf. Process. Syst. **23**, 919–927 (2010)
33. Jiménez, A., Berzal, F., Cubero, J.-C.: Frequent tree pattern mining: a survey. Intell. Data Anal. **14**(6), 603–622 (2010)
34. Keogh, E., Lonardi, S., Chiu, B.Y.-C.: Finding surprising patterns in a time series database in linear time and space. In: KDD, pp. 550–556 (2002)
35. Kleinberg, J.M.: Authoritative sources in a hyperlinked environment. J. ACM **46**(5), 604–632 (1999)
36. Kumar, R., Raghavan, P., Rajagopalan, S., Tomkins, A.: Extracting large-scale knowledge bases from the web. In: VLDB, pp. 639–650 (1999)
37. Latapy, M.: Main-memory triangle computations for very large (sparse (power-law)) graphs. Theor. Comput. Sci. **407**(1–3), 458–473 (2008)
38. Leskovec, J., Backstrom, L., Kleinberg, J.: Meme-tracking and the dynamics of the news cycle. In: KDD, pp. 497–506 (2009)
39. Leskovec, J., Horvitz, E.: Planetary-scale views on a large instant-messaging network. In: WWW, pp. 915–924 (2008)

40. Li, H.-F., Lee, S.Y.: Miningfrequentitemsetsoverdatastreams using efficient window sliding techniques. Expert Syst. Appl. **36**, 1466–1477 (2009)
41. Li, H.-F., Lee, S.Y., Shan, M.-K.: Online mining (recently) maximal frequent itemsets over data streams. In: RIDE-SDMA, pp. 11–18 (2005)
42. Lin, Y.R., Chi, Y., Zhu, S., Sundaram, H., Tseng, B.L.: Facetnet: a framework for analyzing communities and their evolutions in dynamic networks. In: Proceedings of the 17th international conference on World Wide Web, pp. 685–694. ACM (2008)
43. Makino, K., Uno, T.: New algorithms for enumerating all maximal cliques. In: SWAT, pp. 260–272 (2004)
44. Manku, G., Motwani, R.: Approximate frequency counts over data streams. In: VLDB, pp. 346–357 (2002)
45. Mokken, R.J.: Cliques, clubs and clans. Qual. Quant. **13**(2), 161–173 (1979)
46. Moon, J.W., Moser, L.: On cliques in graphs. Isr. J. Math. **3**, 23–28 (1965)
47. Morishita, S.: On classification and regression. In: Discovery Science, pp. 40–57 (1998)
48. Nakamura, Y., Horiike, T., Kuboyama, T., Sakamoto, H.: Extracting research communities from bibliographic data. KES J. **16**(1), 25–34 (2012)
49. Nakano, S., Uno, T.: Efficient generation of rooted trees. Technical report, NII Technical Report NII-2003-005E (2003)
50. Newman, M.E.J.: Fast algorithm for detecting community structure in networks. Phys. Rev. E **69**, 066133 (2004)
51. Newman, M.E.J., Girvan, M.: Finding and evaluating community structure in networks. Phys. Rev. E **69**, 026113 (2004)
52. Nijssen, S., Kok, J.N.: Efficient discovery of frequent unordered trees. In: 1st international workshop on mining graphs, trees, and sequences (MGTS), pp. 55–64 (2003)
53. Nijssen, S., Kok, J.N.: A quickstart in frequent structure mining can make a difference. In: KDD, pp. 647–652 (2004)
54. Oates, T., Cohen, P.R.: Searching for structure in multiple streams of data. In: ICML, pp. 346–354 (1996)
55. Pei, J., Han, J., Mortazavi-Asl, B., Pinto, H., Chen, Q., Dayal, U., Hsu, M.-C.: Prefixspan: mining sequential patterns efficiently by prefix-projected pattern growth. In: ICDE, pp. 215–224 (2001)
56. Pei, J., Han, J., Wang, W.: Constraint-based sequential pattern mining: the pattern-growth methods. J. Intell. Inf. Syst. **28**(2), 133–160 (2007)
57. Qiu, J., Lin, Z.: A framework for exploring organizational structure in dynamic social networks. Decis. Support Syst. **51**(4), 760–771 (2011)
58. Raissi, C., Roncelet, P., Teisseire, M.: SPEED: mining maxirnal sequential patterns over data strearns. In: International IEEE conference on intelligent systems, pp. 546–552 (2006)
59. Schank, T., Wagner, D.: Finding, counting and listing all triangles in large graphs, an experimental study. In: WEA, pp. 606–609 (2005)
60. Seidman, S.B.: Network structure and minimum degree. Social Networks **5**(3), 269–287 (1983)
61. Seidman, S.B., Foster, B.L.: A graph-theoretic generalization of the clique concept. J. Math. Soc. **6**(1), 139–154 (1978)
62. Snowsill, T., Nicart, F., Stefani, M., De Bie, T., Cristianini, N.: Finding surprising patterns in textual data streams. In: International workshop on cognitive information processing, pp. 405–410 (2010)
63. Srikant, R., Agrawal, R.: Mining sequential patterns: generalizations and performance improvements. In: EDBT, pp. 3–17 (1996)
64. Tantipathananandh, C., Berger-Wolf, T., Kempe, D.: A framework for community identification in dynamic social networks. In: Proceedings of the 13th ACM SIGKDD international conference on knowledge discovery and data mining, pp. 717–726. ACM (2007)
65. Tatikonda, S., Parthasarathy, S., Kur, T.M.: Trips and tides: new algorithms for tree mining. In: CIKM, pp. 455–464 (2006)

66. Tomita, E., Tanaka, A., Takahashi, H.: The worst-case time complexity for generating all maximal cliques and computational experiments. Theor. Comput. Sci. **363**(1), 28–42 (2006)
67. Tsukiyama, S., Ide, M., Ariyoshi, H., Shirakawa, I.: A new algorithm for generating all the maximal independent sets. SIAM J. Comput., **6**, 505–517 (1977)
68. Uno, T., Asai, T., Uchida, Y., Arimura, H.: LCM: an efficient algorithm for enumerating frequent closed item sets. In: FIMI (2003)
69. Wang, J., Cheng, J.: Truss decomposition in massive networks. PVLDB **5**(9), 812–823 (2012)
70. Wang, N., Zhang, J., Tan, K.-L., Tung., A.K.H.: On triangulation-based dense neighborhood graph discovery. In VLDB, pp. 58–68 (2010)
71. Wang, N., Zhang, J., Tan, K.L., Tung, A.K.H.: On triangulation-based dense neighborhood graph discovery. Proc. VLDB Endowment **4**(2), 58–68 (2010)
72. Wasserman, S., Faust, K.: Social network analysis: methods and applications. Cambridge University Press, Cambridge (1994)
73. Zaki, M.J.: Efficiently mining frequent trees in a forest. In: KDD, pp. 71–80 (2002)
74. Zaki, M.J.: Efficiently mining frequent embedded unordered trees. Fundam. Inform. **66**(1–2), 33–52 (2005)
75. Zaki, M.J.: Efficiently mining frequent trees in a forest: algorithms and applications. IEEE Trans. Knowl. Data Eng. **17**(8), 1021–1035 (2005)
76. Zaki, M.J., Ogihara, M.: Theoretical foundation of association rules. In: Workshop on data-mining and knowledge discovery (1998)

Source List

Tutorial
Social and Information Network Analysis Course http://www.stanford.edu/class/cs224w/index.html
Material
Stanford Large Network Dataset Collection http://snap.stanford.edu/data
Citation networks http://dblp.uni-trier.de/xml
Internet topology http://topology.eecs.umich.edu/data.html
Youtube http://netsg.cs.sfu.ca/youtubedata
Amazon http://snap.stanford.edu/data/amazon-meta.html
Wikipedia http://users.on.net/~henry/home/wikipedia.htm
Newman's pointers http://www-personal.umich.edu/~mejn/netdata
Mining Program Source
LCM http://research.nii.ac.jp/~uno/code/lcm.html
LCM for sequential mining http://research.nii.ac.jp/~uno/code/lcm_seq.html
FREQT http://research.nii.ac.jp/~uno/code/FREQT_distMay02_j50.tar.gz
Max clique by Makino and Uno http://research.nii.ac.jp/~uno/code/mace.html
Max clique by Tomita et al. http://research.nii.ac.jp/~uno/code/macego10.zip
Social Network Analysis Tools
Gephi http://gephi.org
Network Workbench http://nwb.cns.iu.edu
Pajek http://pajek.imfm.si
igraph http://igraph.sourceforge.net
Others (list of tools) http://en.wikipedia.org/wiki/Social_network_analysis_software

Dominant AHP as Measuring Method of Service Values

Eizo Kinoshita

Abstract This chapter demonstrates the need for promoting service science studies, while touching upon several issues along with a scientific approach to this field, through a perspective of a new paradigm for the twenty-first century. The author's discussion postulates that identifying and understanding the characteristics as well as the essence of services is vital, while breaking them down into fundamental elements, which will eventually be combined and recreated into an innovative form. The author believes that primary role of service science is to provide new values, which are to be co-created through human efforts facilitated by advanced technologies, and that analyzing and interpreting the characteristics and the essence of such values is significant. Furthermore, establishing a language and concepts (including mathematical concepts) to be shared by those who engage in this study, as well as a social common ground for service operations is important in order to have the above mentioned vision permeate into a society. Then, the author proves that computer science, a paradigm of the twentieth century, is being replaced by a new paradigm which centers on service science in the new millennium, by focusing on such concepts as a network society and a new type of capitalism. Presently, however, service science lacks theoretical methods for measuring service values, which has mostly been done based on human experiences and experience-based intuition. Therefore, it is necessary to create a service model, which will enable the measurement of values of all different types of services, including those provided by public administrations, corporations, as well as by non-profit organizations (NPOs), which leads the author to insist on the need for establishing a measuring method for service values. And in the process of finding such a method, the author discusses the merits and defects inherent in utility theory, a measuring method of economic value of goods in an industrialized society, proposed by John von Neumann, which the author believes is surpassed by Analytic Hierarchy Process (AHP), a measuring method of the values of goods in a

E. Kinoshita (✉)
School of Urban Science, Meijo University, Tokyo, Japan
e-mail: kinoshit@urban.meijo-u.ac.jp

G. A. Tsihrintzis et al. (eds.), *Multimedia Services in Intelligent Environments*,
Smart Innovation, Systems and Technologies 24, DOI: 10.1007/978-3-319-00372-6_8,
© Springer International Publishing Switzerland 2013

service economy, proposed by Thomas L. Saaty. This is because a utility function which represents values is expressed based on an absolute scale (suitable for an industrialized society) in utility theory, whereas in AHP, it is expressed based on a relative scale (suitable for a service economy). However, the author proves mathematically a fundamental flaw of Saaty's AHP in the chapter, and demonstrates that Dominant AHP, proposed by Kinoshita and Nakanishi, is superior to Saaty's AHP as a measuring method of service values. Finally, the chapter shows that the establishment of a measuring method of service values would help facilitate innovations for the new millennium, while fostering grand designs for economic growth, which the author believes would ultimately enable the construction of a service science society.

1 Introduction

Given that a paradigm for the twentieth century was computer science, it could be said that a paradigm for the new millennium is service science (See Table 1).

What the author would like to emphasize in Table 1 is the fact that a concept of service science has become a paradigm of the new millennium.

There are several principles and theorems, which the author believes demonstrate the characteristics of a hierarchical society (computer science), where conventional capitalism is supported, as well as those of a network society, where a new form of capitalism (a service science society) is supported, shown below.

[Principles] (the spirit of capitalism valid both for a hierarchical society and a network society)

The following three points were indicated by Max Weber:

(1) High-level ethics (soberness, faithfulness, honesty, reliance, time is money, etc.)
(2) Rational purposes (corporations tend to seek maximization of profits and consumers tend to seek to maximize their own utilities)

Table 1 Paradigms in twentieth and twenty-first centuries

	Twentieth century	Twenty-first century
Paradigm	Computer science (the end of nineteenth century to early twentieth century)	Service science (the end of twentieth century to early twenty-first century)
Place	Hungary, Budapest (Cafe New York)	America, New York (IBM)
Key person	John von Neumann	Bill Gates
Society	Hierarchical society	Network society
Capitalism	Conventional capitalism	New capitalism
Prophet	Friedrich Wilhelm Nietzsche	Alvin Toffler

(3) Approval of interests (Approval of interests means to recognize corporate tendency to seek profits as something good).

Capitalism came into existence with the formation of the above-mentioned concepts. And it was in this capitalist society that a hierarchical society was formed, along with a network society which is currently in the process of being formulated. Thus, it can be said that the spirit of capitalism as indicated by Max Weber is vital for sustaining both the conventional capitalism, as well as a new form of capitalism.

Theorem 1 (Rules of a hierarchical society) *A hierarchical society helps people to lead a stable life, and to achieve well being, unless they revolt against it. The formation of capitalist concepts facilitates the society to create more and more capitalists as well as workers. This society, however, is not fully liberalized.*

Theorem 2 (Rules of a network society) *In a network society, people are given a tremendous amount of freedom, along with numerous new business chances. However, it would not be able to produce more capitalists or workers who are suitable for the capitalist way of things, unless the society continues to cherish the spirit of capitalism. In other words, without the capitalist spirit, the well being of people would eventually be lost.*

In the meantime, the analysis and interpretation of the characteristics and properties inherent in service science is necessary, since it provides new values, which are co-created through human efforts facilitated by advanced technologies. A common social ground for analyzing and interpreting services is also required so that service science can be broadly accepted in a human society. As to the measurement of service values in this field of study, however, human experiences and experience-based intuition have played a large role, without any theoretical background. This naturally led the author to insist on the need for a method to measure service values, based on a service model, because such a method would enable the measurement of service values in diversified areas (administrations, firms and NPOs).

In describing how beneficial Dominant AHP (Analytical Hierarchy Process) is as a method to measure the economic values of goods in a service society, the author, in this chapter, touches upon the significance of service science, explains its approach, then proceeds to analyze Saaty's AHP and Dominant AHP as measuring methods of service values. The author goes on to describe Saaty's AHP and Dominant AHP from a perspective of utility functions of utility theory, a measuring method of the economic values of goods in an industrialized society, proposed by John von Neumann. In this utility theory, utility functions which denote values are expressed based on an absolute scale (which is applicable to an industrialized society, based on labor theory of value), whereas in AHP, they are expressed based on a relative scale (applicable to an industrial society, which is not based on labor theory of value).

However, there is a problem of reversal of ranks of alternatives in respect to Saaty's AHP. This leads the author to prove how useful Dominant AHP, proposed by Kinoshita and Nakanishi, is because it is free from such a shortcoming by comparing its aspect as a measuring method of service values with an expressive form of utility functions.

The chapter consists of Sect. 2, which explains the significance of service science and its approach; Sect. 3, in which Saaty's AHP and Dominant AHP as measuring methods of service values are described mathematically; Sect. 4, which compares expressive forms of Saaty's AHP and Dominant AHP from a viewpoint of utility functions; as well as Sect. 5, a conclusion.

2 Necessity of Measuring Service Values

2.1 Significance of Service Science

In this section, the author describes how significant service science is, based on the argument in the Ref. [1]. In Japan as well as in other developed nations, almost 70 percent of the workforce belong to the service sector, while service industries occupy nearly 75 percent of the gross domestic product (GDP) in each country. Service industries have become the main engine of economic growth in those countries since 2001 (after the turn of a new millennium).

Therefore, there is a need for a dramatic improvement of services in the public sector (central and local governments), the private sector (private firms), also in the non-profit sector (universities, hospitals and so forth) in terms of efficiency and productivity. Furthermore, it is important not only to develop the tertiary sector (service sector), but also to promote service-science-related properties in primary sector (agriculture and fisheries) and secondary sector (manufacturing). And the author believes that innovation in services is vital in order to enhance the competitiveness of these countries for this end.

It has been said that services are "intangible" because "they cannot be looked at and touched, having no physical substance," and that they are "simultaneous" because "they are produced and consumed at the same time." The "intangible" and "simultaneous" qualities of services are also connected with their "variability" and "perishability." That is to say, services are variable because "there could be differences in the quality of services," and are perishable because "services cannot be stored."

Thus, it is obvious that the measurement of service values is important, and that the establishment of such a method is among the priorities in service science. By establishing a measuring method of various service values, it would be possible to analyze, compare and understand diversified qualities and attributes of services.

2.2 Scientific Approach to Service Science

Conventionally, the measurement of service values in the area of service science depends largely on human experiences and experience-based intuition, lacking any theoretical basis. Thus, it is necessary to establish such a model in the area of service science as to be applied to the measurement of service values. With such a model, it would be possible to measure values in all the areas of services (including administration, firms, NPOs and so forth).

A scientific approach is also necessary, in order to establish "service science" as scholarship which aims at examining services from scientific points of view, as are shown in the following by the author [1].

① It should start with the discovery and appreciation of the characteristics and attributes of services. It is also necessary to bring out a new innovation by resolving and synthesizing those characteristics and attributes, discovered and understood in service science.

② Service science provides a society with new values, which are co-created through human efforts and advanced technologies, and it is important to analyze and interpret the characteristics and attributes discovered in the process described in ①.

③ A common language and concepts (including mathematics and concepts) for service operation, along with a common social ground are vital so that service science can take root in a human society.

Because service science is applicable to diversified areas of services including those provided by public organizations, private companies as well as by NPOs, a common method to measure the values of those services is necessary.

Furthermore, the author classifies services into three categories, or "stock service," "flow service" and "flow rate of change service" from a scientific point of view. Physics, which was established during the seventeenth century, for instance, has matured as such concepts like "distance," "velocity," "acceleration" and "growth rate" came to be formulated. Therefore, in order to establish service science as scholarship, it is significant to review the development process of "modern physics" and "modern economics" from a viewpoint of services, while in the process of classifying services into those categories. "Stock service" is a concept similar to that of distance in physics, and that of assets in economics; "flow service" is a concept similar to that of velocity in physics, and that of income in economics; "flow rate of change service" is a concept similar to that of acceleration in physics, and that of growth rate in economics. "Stock service" signifies services relating to infrastructure and institutions, including social and information-related infrastructure, provided mainly by public sector (administration). "Flow service" includes services provided generally and on a daily basis, mostly in public and private sectors. And "flow rate of change service" is something provided on special occasions in private sector. Table 2 shows the three categories of services (Kinoshita [1]).

Table 2 Categories of services

Mathematics	Physics	Economics	Service science	Example of services	Angle of measuring service values
Original variable	Distance	Assets (stock)	Stock service	Administrative services Social security system Social infrastructure Information infrastructure	Time-axis integration: cost-benefit analysis
Differentiate once with respect to time	Velocity	Income (Flow)	Flow service (daily service)	Fast food Supermarkets Starbucks, etc.	Time-axis differentiation: CS research
Differentiate twice with respect to time	Acceleration	Growth rate (Percentage change of flow)	Service in percentage change of flow (non-daily service)	Kagaya Ritz-Carlton, Osaka Financial engineering, etc	Comfortable change: Fractal measurement

(Cited from Kinoshita [1], p. 27)

3 AHP as a Measuring Method of Service Values

3.1 Saaty's AHP

In this section, the author explains conventional AHP, a decision-making method proposed by Saaty [2], from a viewpoint of a service-value measuring method. In Saaty's AHP, a hierarchical structure is formed to express the relationship between such elements as general purposes (Goal), criteria of evaluation (Criteria) and alternatives (Alternative), in respect to a problem which requires a resolution and/ or a decision making.

The (Goal), consisting of a single element, is placed at the uppermost level of this hierarchical structure, and (Criteria), or criteria of evaluation for decision making, is set at the secondary and the lower level. It is up to a decision maker to decide how many (Criteria) elements are to be included on this level. A hierarchical structure needs to be built in a way as to reflect the reality of a problem generated under different circumstances, for its resolution. The building of a structure is completed by placing (Alternative) at its lowest level.

Next, a pairwise comparison is to be repeated for all the elements on every level to define the weight of each of them. In conducting a pairwise comparison, two elements which belong to the same level are selected, and are compared to decide which of the two is more significant. The process is conducted on (Criteria) elements in terms of (Goal), also on (Alternative) elements in terms of (Criteria). Suppose n denotes the number of elements to be compared, a decision maker must compare $n(n-1)/2$ pairs. Table 3 demonstrates the criteria of significance to be applied when comparing two elements.

The results, obtained by applying the criteria shown in Table 3 to the results, obtained from pairwise comparison, could form pairwise-comparison matrix A. The constituents of its main eigenvector represent the weight of respective elements. When creating a pairwise-comparison matrix, by conducting pairwise comparison on the elements included in each level, it could bring the following weight vector, which has n (n denotes the number of elements) worth of weights as constituents.

Table 3 Criteria of significance and definition

Criteria of significance	Definition
1	Equal importance
3	Weak importance
5	Strong importance
7	Very strong importance
9	Absolute importance

(2, 4, 6, 8 are used as a supplementary as mean value)

$$W = \begin{bmatrix} w_1 \\ w_2 \\ \vdots \\ w_n \end{bmatrix} \tag{1}$$

Suppose that a_{ij} denotes the degree of significance of element i in terms of j, it would bring about pairwise comparison matrix $A = [a_{ij}]$. And when weight W is already obtained, $A = [a_{ij}]$ can be as below.

$$A = \begin{bmatrix} w_1/w_1 & w_1/w_2 & \cdots & w_1/w_n \\ w_2/w_1 & w_2/w_2 & \cdots & w_2/w_n \\ \vdots & \vdots & \cdots & \vdots \\ w_n/w_1 & w_n/w_2 & & w_n/w_n \end{bmatrix} \tag{2}$$

However, $a_{ij} = w_i/w_j$, $a_{ij} = 1/a_{ij}$, $i, j = 1, 2,..., n$. Also the following is valid for i, j, and k.

$$a_{ij} \times a_{jk} = a_{ik} \tag{3}$$

The equation demonstrates a state where a decision maker's judgements are perfectly consistent. When this pairwise comparison matrix A is multiplied by W, a weight vector, vector nW is obtained as is shown in Eq. (4).

$$AW = nW \tag{4}$$

This equation can be transformed into the following equation which expresses eigenvalue.

$$(A - nI)W = 0 \tag{5}$$

To bring $W \neq 0$ valid in this equation, n needs to be eigenvalue of A, and if that is in the case, W is eigenvector of A. And since the rank of A is 1, eigenvalue $\lambda_i (i = 1, 2, ..., n)$ turns out to be zero in all the cases except for one. And the total of A's elements of the main diagonal is denoted by n. The only λ_i, which is not zero, is represented by n, and weight vector W turns out to be a normalized eigenvector in terms of A's biggest eigenvalue. However, under actual circumstances, W is unknown and should be obtained through pairwise comparison matrix A, after it is applied to an actual situation. Suppose that A's biggest eigenvalue is denoted by λ_{max}, it turns out to be as follows.

$$AW = \lambda_{max}W \tag{6}$$

By applying this equation, W can be obtained. In other words, eigenvector in terms of the biggest eigenvalue λ_{max} of a pairwise comparison matrix turns out to be the weight of respective criterion. In fact, when applying AHP to an actual situation, other than a method to calculate eigenvalue of a pairwise comparison matrix, weights can also be obtained by utilizing geometric mean of rows in a pairwise comparison matrix.

Suppose that W denotes a weight of an evaluation criterion which is immediately above the weight of a certain alternative, and when calculating the weight of respective alternative in terms of the criteria originally belonging to those alternatives, it leads to bringing about evaluation matrix M. Suppose $E_1, E_2, ..., E_n$ denote total evaluation values, which express degree of relative priority for each alternative, a vector comprising them as constituents turns out to be Eq. 7.

$$E = \begin{bmatrix} E_1 \\ E_2 \\ \vdots \\ E_n \end{bmatrix} = MW \tag{7}$$

3.2 Dominant AHP

In this section, the author describes Dominant AHP [3, 4] proposed by Kinoshita and Nakanishi. While in Saaty's AHP, the values of all the alternatives are normalized to 1, Dominant AHP focuses on a certain alternative, makes it a criterion and conducts an evaluation.

In Dominant AHP, such an alternative, which is made into a benchmark of evaluation, is called a dominant alternative, and the rest of other alternatives are called subordinate alternatives.

Suppose that X, Y and Z are dominant alternatives, and that their weights of criteria are denoted by W_x, W_y, W_z, respectively, a matrix comprising them as constituents is shown as follows.

$$W = \left(W_x, W_y, W_z \right) \tag{8}$$

Matrix M and matrix M_i comprising constituents, or evaluation values under criteria, are expressed by Eqs. 9 and 10, respectively.

$$M = \begin{bmatrix} a_{XI} & a_{XII} \\ a_{YI} & a_{YII} \\ a_{ZI} & a_{ZII} \end{bmatrix} \tag{9}$$

$$M_i = M \begin{bmatrix} 1/a_{iI} & 0 \\ 0 & 1/a_{iII} \end{bmatrix} = MA_i^{-1} \tag{10}$$

However, in Eq. 10, A_i should be as follows.

$$A_i = \begin{bmatrix} a_{iI} & 0 \\ 0 & a_{iII} \end{bmatrix} \tag{11}$$

>156 E. Kinoshita

Dominant AHP is a method where a particular alternative is selected and made a criterion to evaluate other alternatives. Specifically, evaluations are conducted in respect to subordinate alternatives, after the evaluation values of dominant alternatives under criteria are normalized to 1.

For example, when alternative X is selected as a dominant alternative, the weight of criteria in terms of alternative X is expressed by Eq. 12.

$$A_X A_X^{-1} W_X = W_X \tag{12}$$

Evaluation values under criteria is expressed by Eq. 13, in which the evaluation value of alternative X is normalized to 1.

$$M = M A_X^{-1} \tag{13}$$

Aggregate score, which expresses degrees of relative priority of alternatives, is expressed by Eq. 14.

$$M_X \left(A_X A_X^{-1} W_X \right) = M A_X^{-1} W_X \tag{14}$$

Estimate values of the weight of criteria in terms of alternative Y can be expressed by $A_Y A_X^{-1} W_X$, bringing aggregate score to be expressed by Eq. 15.

$$M_Y \left(A_Y A_X^{-1} W_X \right) = M A_X^{-1} W_X \tag{15}$$

By the same token, estimate values of the weight of criteria in terms of alternative Z can be expressed by $A_Z A_X^{-1} W_X$, bringing aggregate score to be expressed by Eq. 16.

$$M_Z \left(A_Z A_X^{-1} W_X \right) = M A_X^{-1} W_X \tag{16}$$

Thus, Eqs. 14, 15 and 16 prove that Dominant AHP owns different aggregate scores depending on each alternative, and when normalizing all those equations so that the total should be 1, the aggregate score of alternative X, and that of alternative Y as well as that of alternative Z are all identical to one another.

In Saaty's AHP, decision making is conducted in such a manner as to bring everything, from alternatives to a goal, to flow in the same direction. Furthermore, all the alternatives are treated as equal, and the evaluation values of alternatives are normalized so that the total should be 1. Saaty's AHP, however, has the following problems.

① Because the aggregate score of evaluation values of alternatives are normalized to 1 in Saaty's AHP, it sometimes causes a problem of reversal of ranks. Belton and Gear demonstrate such examples concerning this [5].

② When people make decisions and conduct evaluations in actual circumstances, they often select a certain alternative, make it a benchmark or a tentative criterion, compare it with the rest of other alternatives to conduct evaluations. Or in other cases, some select a certain criterion to evaluate alternatives. Saaty's AHP, however, is not applicable to either case.

Whereas in Dominant AHP, people make decisions by focusing on a certain alternative (dominant alternative) and make it a criterion for evaluation. In other words, an alternative with distinctive characteristic is selected and made into a criterion, instead of treating all the alternatives on an equal basis, when making a decision. Also in this method, a particular alternative is made as a dominant alternative, and evaluation weights are defined based on the evaluation of this dominant alternative, by normalizing it to 1. Although weights of criteria differ in case a different alternative is selected and made into a dominant alternative, the aggregate score of alternatives is always consistent.

In fact, because there is no need to normalize each evaluation value in Dominant AHP, it is free from reversal of ranks, which is inherent in Saaty's AHP. Furthermore, because in Dominant AHP, a specific dominant alternative is selected depending on a type of decision making required under different circumstances, it is applicable to a wider range of decision making processes. Kanda et al. applied Saaty's AHP and Dominant AHP, as methods to measure service values, to the quantitative evaluation of food texture [6].

4 AHP and Dominant AHP from a Perspective of Utility Function

4.1 Expressive form of Multi-Attribute Utility Function

Multi-attribute utility functions are among methods which have been employed for measuring the values of economic goods. They are applicable to those goods with multiple attributes, such as x_i, $i = 1, 2, \ldots m$, to which a common measurement is often not applicable in conducting evaluations. Thus, multi-attribute utility functions tend to have a complex form. And this complexity could spoil their utility when employing them to conduct evaluations of practical issues. Therefore, the author believe that it is necessary to discover such a functional form of muti-attribute utility functions as to be expressed in a plain and easy-to-handle form, without losing properties appropriate for analyzing mutipurpose systems.

One such form is an additive form as is shown in Eq. 17 [7].

$$u(x) = \sum_{i=1}^{m} k_j u(x_i), \quad \sum_{i=1}^{m} k_i = 1, 0 \le u(x_i) \le 1 \tag{17}$$

As is shown in this equation, utility is acquired by adding all the products obtained by weight k_i multiplied by utility value $u(x_i)$ for each attribute.

When actually measuring a utility function, it is necessary to set up the range from best to worst for each attribute in order to clarify the preferences of an individual, such as averting dangers, seeking dangers and/or none of those. It is possible to bring about a utility function mathematically; however, applying it in actual circumstances is not easy.

4.2 Saaty's AHP from a perspective of utility function

As is explained in Sect. 3, AHP, enabling the prioritization of quantitative alternatives, is useful as a method which can be applied to the measurement of service values. However, this doesn't necessarily mean that any AHP models, including Saaty's AHP and Dominant AHP, are applicable to the measurement of service values, because there are differences among them.

In this clause, the author describes the characteristics of Saaty's AHP, by comparing it with expressive forms of utility functions.

① In Saaty's AHP, the evaluation of multiple alternatives is conducted by repeating a pairwise comparison for all the alternatives, based on a pairwise comparison matrix, instead of conducting evaluations of all the alternatives at once.

② Eigenvector in terms of the largest eigenvalue of the pairwise comparison matrix in ① brings about evaluation values. Because scalar multiplication can be made at random concerning eigenvector, it can be normalized so that the total of attributes should be 1. Geometric mean of a pairwise comparison matrix can also be utilized.

③ In case there are several criteria, the weight of each criterion is calculated by repeating a pairwise comparison among the criteria.

④ Aggregate score of each alternative can be obtained by ② and ③.

In other words, alternatives and criteria of Saaty's AHP can be expressed as follows.

Suppose m worth of criteria are represented by C_1, C_2 ..., C_i..., $C_m (i = 1, 2, ...,m)$, and n worth of alternatives are expressed by $A_1, A_2 ..., A_j..., A_n (j = 1, 2,..., n)$. Also, m worth of weights of criteria can be expressed by $k_1, k_2..., k_m$, and suppose that the evaluation value of n worth of alternatives under k_i are represented, respectively, by $w_{i1}, w_{i2}..., w_{im}$, aggregate score is expressed by Formula (18).

$$U_i = \sum_{i=1}^{m} k_i w_{ij} \tag{18}$$

In other words, the aggregate score represents a utility value, which signifies the degree of relative priority in respect to alternative j. When comparing Eqs. 17 and 18, n worth of alternatives under criterion i by applying AHP, or A_1, A_2 ..., A_j ..., A_n, and the respective evaluation value $w_{i1}, w_{i2}..., w_{in}$, express utility values evaluated in terms of n worth of alternatives, or $u_i(x_{i1})$, $u_i(x_{i2})..., u_i(x_{in})$. Here, $u_i(x_{ij})$, $j = 1,2,...,n$ denote the evaluation value of alternative j in terms of criterion (attribute) i. Thus, it can be said that the aggregate significance of an alternative evaluated by AHP expresses a multi-attribute utility value in respect to several criteria.

In Saaty's AHP, Eqs. 19 and 20 are normalized. As to Eq. 19, the sum of the weights of criteria is normalized to 1.

$$\sum_{i=1}^{m} k_i = 1 \tag{19}$$

As to Eq. 20, the sum of the evaluation value of respective alternative under a certain criterion is normalized to 1.

$$\sum_{i=1}^{n} w_{ij} = 1 \tag{20}$$

There is no problem with the normalization of Eq. 19, which is consistent with the normalization of the sum of weights $k_1, k_2, \ldots, k_m$, of Eq. 17 so that it should be 1. However, the normalization of Eq. 20 in terms of alternatives corresponds with the normalization of the sum of evaluated utility values $u_i(x_{i1}), u_i(x_{i2}) \ldots, u_i(x_{in})$, so that it should be 1, in terms of n worth of alternatives in utility theory. However, it is irrelevant from a perspective of utility theory. Because in utility theory, $u_i(x_{ij})$ of Eq. 17 is unique within the range of positive linear transformation, its value is normalized to zero in terms of the worst attributive value, and to 1 in terms of the best attributive value. Thus the normalization of Eq. 20 results in rank reversal of aggregate evaluation score [5], which demonstrates a fallacy of Saaty's AHP.

4.3 Dominant AHP from a viewpoint of utility function

In this clause, the author describes an expressive form of Dominant AHP. The procedure of Dominant AHP has already been described mathematically in Sect. 3. When comparing it with an expressive form of utility functions, the characteristics of Dominant AHP can be summarized as follows.

① In Dominant AHP, criteria and alternatives are grouped into ranks. A dominant alternative is selected as a tentative benchmark for evaluation.
② A pairwise comparison, which is conducted on alternatives, aims at evaluating the dominant alternative selected in ①. Respective weight of criteria is brought about through the biggest eigenvalue of a pairwise comparison matrix in terms of eigenvector, and the aggregate score of weights is normalized to 1.
③ The evaluations of alternatives under respective criterion are brought about by applying a pairwise comparison method for all the alternatives. In other words, the ratio of evaluation of respective alternative can be obtained through eigenvector in terms of the biggest eigenvalue of a pairwise comparison matrix, and the evaluation value of a dominant alternative is normalized so that it should be 1.
④ Aggregate score U_i can be obtained through ② and ③. However, the aggregate score of a dominant alternative turns out to be 1.

Through the process of ① through ④, aggregate score obtained based on Dominant AHP can be expressed as follows.

$$U_j = \sum_{i=1}^{m} k_i w_{ij} \tag{21}$$

However, as is explained in Sect. 3, the normalization of criteria is expressed as $\sum_{i=1}^{m} k_i = 1$, and the normalization of aggregate score can be demonstrated as $U_{j*} = \sum_{i=1}^{m} k_i w_{ij*} = 1$.

This demonstrates that Eq. 17, which is expressed by an additive form of utility functions, and Eq. 21, which expresses Dominant AHP, are structurally the same. Thus, it can be said that Dominant AHP is a type of an additive form of muti-attributive utility functions, and that rank reversal as is shown by Belton and Gear et al. should not appear when applying Dominant AHP.

Thus, it is able to state that as an AHP method to measure service values, Dominant AHP is more useful than Saaty's AHP.

5 Conclusion

In this chapter, the author analyzed Saaty's AHP and Dominant AHP, while insisting on the need for a theoretical method to measure service values in the area of service science. This is because conventionally, the measurement of values in service science has depended largely on human experiences and experience-based intuition, which is quite uncertain as a method. In the past, in utility theory, a method to measure the value of goods, utility functions, which denote values, have been expressed on an absolute scale beginning from zero to 1, whereas in AHP, values are expressed based on a relative scale. The author then discussed that Dominant AHP is more useful than an expressive form of utility functions, because it is free of a rank reversal problem, and that it is applicable to a wider range of decision-making processes. As is shown in Ref. [6], there are actual cases where Dominant AHP was applied in the measurement of service values. Thus, the author believes that it is his future task to apply Dominant AHP to a wider range of analyses of service science, and to prove its validity through practical cases.

References

1. Kinoshita, E.: Theory and practice of service science, Kindai Kagaku-sha, 11–28. http://www.kindaikagaku.co.jp (2011)
2. Saaty, T.L.: The Analytic Hierarchy Process, McGraw-Hill http://www.mcgraw-hill.com/ (1980)
3. Kinoshita, E., Nakanishi, M.: Proposition of a new viewpoint in AHP, Jpn. Soc. Civ. Eng. J. 569/IV-36:1–8. http://library.jsce.or.jp/jsce/open/00037/1997/mg03.htm#569 (1997)

4. Kinoshita, E., Nakanishi, M.: Proposal of new AHP model in light of dominant relationship among alternatives, J. Oper. Res. Soc. Japan, **42**(2), 180–197. http://www.orsj.or.jp/~oldweb/e-Library/jorsj42.htm#vol42b (1999)
5. Belton, V., Gear, T.: On a short-coming of Saaty's method of analytic hierarchies, Omega, **11**(3), 228–230. http://www.sciencedirect.com/science/journal/03050483/11/3 (1982)
6. Kanda, T., Sugiura, S., Kinoshita, E.: Food texture evaluation by dominant AHP for the promotion of productivity of services in food industries. Kansei Eng. Int. J. **10**(2), 95–100. https://www.jstage.jst.go.jp/browse/jjske/10/2/_contents/-char/ja/ (2011)
7. Seo, F.: Technology of ideas–business decision making under ambiguous environment. Yuhikaku. http://www.yuhikaku.co.jp (1994)

Applications of a Stochastic Model in Supporting Intelligent Multimedia Systems and Educational Processes

Constantinos T. Artikis and Panagiotis T. Artikis

Abstract It is generally recognized that stochastic models constitute extremely strong analytical tools for a very wide variety of fundamental research areas of modern informatics. The theoretical contribution of the present chapter consists of the formulation and investigation of a stochastic model being an extension of the concept of minimum of a random number of nonnegative random variables. Moreover, the practical contribution of the present chapter consists of the establishment of applications of the formulated stochastic model in the development, assessment, and implementation of operations for supporting intelligent multimedia systems and educational processes.

1 Introduction

Information risk management is recognized as the process of manipulating risks threatening an information system, or equivalently information risks, according to the public interest, human safety, environmental issues, and the law. It contains the planning, organizing, directing, and controlling operations implemented with the purpose of developing an efficient plan that impairs the negative effects of information risks [1–5]. Information risk management accepts the general principles of risk management and uses them to the integrity, availability, and confidentiality of information assets and the information environment. It is readily

C. T. Artikis (✉)
Ethniki Asfalistiki Insurance Company, 103-105, Syngrou Avenue,
117 45 Athens, Greece
e-mail: ctartikis@gmail.com

P. T. Artikis
Section of Mathematics and Informatics, Pedagogical Department,
University of Athens, 20 Hippocratous Str, 106 80 Athens, Greece
e-mail: ptartikis@gmail.com

G. A. Tsihrintzis et al. (eds.), *Multimedia Services in Intelligent Environments*,
Smart Innovation, Systems and Technologies 24, DOI: 10.1007/978-3-319-00372-6_9,
© Springer International Publishing Switzerland 2013

understood that information risk management must be included into decision making for usual activities and if appropriately applied can be considered as a strong tool for handling information risks in a proactive rather than in a reactive way. Moreover, it is easily understood that effective information risk management procedures must assist and tend to the total security, culture, operational, and business activities of organizations.

The decisions supporting information risk management have to be compatible with operational and strategic goals and priorities of organizations. In addition, organizations have to identify fundamental elements that may strengthen or weaken their skill to handle information in a secure way. A systematic and logical consideration of information risk management is required independently of the assessment of risk for the implementation of a significant project, for ordinary operational controls and procedures, or the implementation of new information standards. If it is implemented in an appropriate way, information risk management process will not only determine and keep causes of risks from influencing the behaviour of organizations, but also determine opportunities that may offer gain or advantage. Confirming commitment to the process and performance of information risk management, the process can also be formally connected to the outcomes and performance measures of organizations. Successful implementation of information risk management implies that the responsibilities for accomplishing tasks and monitoring risks have to be precisely described. Coordination of information risk management falls across all sections of organizations, and all staff undertake some responsibility for handling risks in their business environments. Resourcing requirements for implementing, monitoring and modifying information risk strategies, should be considered as part of planning for information security and management procedures.

The stochastic character of severity, frequency, duration and timing of information risk makes necessary the use of stochastic models in the field of information risk management practices. During the last three decades, the growth of information risk management was very significant [6]. In particular, the basic information risk management operations are considered more skilful by formulating and using stochastic models for describing, analyzing and solving difficult problems of information systems. An entire examination of the contribution of stochastic modelling to the advancement of information risk management, as a modern organizational discipline of exceptional practical importance, makes quite clear an ongoing significance attributed to the use of probability distributions theory in measurement, evaluation and treatment operations for information risks. The present chapter mainly concentrates on extending the theoretical and practical applicability of stochastic models for supporting fundamental systems and processes. More precisely, the chapter formulates a stochastic model suitable for investigating risks threatening a category of information systems and assessing certain processes.

The following three purposes are implemented by the present chapter. The first purpose is the formulation of a stochastic model by making use of a discrete random variable, a sequence of discrete random variables and a sequence of nonnegative random variables. The second purpose is the establishment of sufficient conditions for evaluating the distribution function of the formulated stochastic model. The establishment of applications of the formulated stochastic model, in making decisions related to the support of intelligent multimedia systems and educational processes, constitute the third purpose.

2 Formulating a Minimum of a Random Number of Nonnegative Random Variables

The concept of maximum of a random number of nonnegative random variables, the concept of minimum of a random number of nonnegative random variables and the concept of random sum of nonnegative random variables constitute very strong analytical tools of probability theory. Moreover, these concepts are extremely useful for formulating stochastic models with significant applications in economics, management, systemics, informatics, logistics, engineering, operational research, biology, meteorology, cindynics, insurance, and other fundamental disciplines. Recently, properties and practical applications of the concept of maximum of a random number of nonnegative random variables, incorporating the concept of random sum of discrete random variables, have been established by Artikis and Artikis [7]. The present chapter extends the results of these authors. More precisely, the chapter concentrates on the formulation and investigation of a stochastic model which is an extension of the concept of minimum of a random number of nonnegative random variables. The constituent elements of the formulated stochastic model are a discrete random variable, a sequence of discrete random variables, and a sequence of nonnegative random variables.

We suppose that

$$N$$

is a discrete random variable with values in the set

$$\mathbf{N} = \{1, 2, \ldots\}$$

and we also suppose that

$$\{S_n, n = 1, 2, \ldots\}$$

is a sequence of discrete random variables with values in the set

$$\mathbf{N} = \{1, 2, \ldots\}.$$

We consider the random sum

$$R = S_1 + S_2 + \ldots + S_N.$$

Moreover, we suppose that

$$\{X_r, r = 1, 2, \ldots\}$$

is a sequence of nonnegative random variables. We consider the stochastic model

$$T = \min\{X_1, X_2, \ldots, X_R\}.$$

The present chapter mainly concentrates on the establishment of theoretical properties and practical applications, in some significant disciplines, of the above minimum of a random number of nonnegative random variables.

It is of some particular importance to mention that the incorporation of the fundamental discrete random sum

$$R = S_1 + S_2 + \cdots + S_N$$

in the structure of the stochastic model

$$T = \min\{X_1, X_2, \ldots, X_R\}$$

substantially supports the applicability in theory and practice of such a stochastic model.

3 Distribution Function of the Formulated Minimum

The present section of the chapter is devoted to the establishment of sufficient conditions for evaluating the distribution function of the minimum of a random number of nonnegative random variables formulated by the previous section. It is easily understood that the evaluation of this distribution function provides designers and analysts of systems with valuable information for investigating the applicability of the corresponding stochastic model in different practical disciplines. Below, we follow Artikis and Artikis [7] in order to establish sufficient conditions for evaluating the distribution function of the minimum of a random number of nonnegative random variables formulated by the second section of the chapter.

Theorem *We suppose that the discrete random variable*

$$N$$

has probability generating function

$$P_N(z), \tag{1}$$

the discrete random variables of the sequence

$$\{S_n, n = 1, 2, \ldots\}$$

are independent and distributed as the random variable

$$S$$

with probability generating function

$$P_S(z) \qquad\qquad (2)$$

and the nonnegative random variables of the sequence

$$\{X_r, r = 1, 2, \ldots\}$$

are independent and distributed as the random variable

$$X$$

with distribution function

$$F_X(x). \qquad\qquad (3)$$

We consider the random sum

$$R = S_1 + S_2 + \ldots + S_N$$

and we suppose that

$$N, \{S_n, n = 1, 2, \ldots\}, \{X_r, r = 1, 2, \ldots\}$$

are independent then the distribution function of the stochastic model

$$T = \min\{X_1, X_2, \ldots, X_R\}$$

has the form

$$F_T(t) = 1 - P_N(P_S(1 - F_X(t))).$$

Proof From the above assumptions it follows that

$$N, \{S_n, n = 1, 2, \ldots\}$$

are also independent. Hence from (1) and (2) it is readily shown that the probability generating function of the random sum

$$R = S_1 + S_2 + \ldots + S_N$$

has the form

$$P_R(z) = P_N(P_S(z)). \qquad\qquad (4)$$

Moreover, from the same assumptions it can be proved that

$$R = S_1 + S_2 + \ldots + S_N, \{X_r, r = 1, 2, \ldots\}$$

are also independent. Hence from (3) and (4) it is easily shown that the distribution function of the stochastic model

$$T = \min\{X_1, X_2, \ldots, X_R\}$$

has the form

$$F_T(t) = 1 - P_N(P_S(1 - F_X(t))). \tag{5}$$

Two particular cases with some theoretical and practical importance of the above distribution function are the following.

First, we suppose that the discrete random variable

$$N$$

follows the Sibuya distribution with probability generating function

$$P_N(z) = 1 - (1 - z)^a, 0 < a \le 1, \tag{6}$$

and the discrete random variable

$$S$$

follows the Sibuya distribution with probability generating function

$$P_S(z) = 1 - (1 - z)^\gamma, 0 < \gamma \le 1. \tag{7}$$

From (4), (6) and (7) it is easily seen that the random sum

$$R = S_1 + S_2 + \ldots + S_N$$

follows the Sibuya distribution with probability generating function

$$P_R(z) = 1 - (1 - z)^\delta, 0 < \delta \le 1, \tag{8}$$

where

$$\delta = a\gamma.$$

Moreover, we suppose that the nonnegative random variable

$$X$$

follows the Beta distribution with distribution function

$$F_X(x) = x^\kappa, 0 \le x \le 1, \kappa > 0. \tag{9}$$

From (5), (8) and (9) it is easily seen that the stochastic model

$$T = \min\{X_1, X_2, \ldots, X_R\}$$

follows the Beta distribution with distribution function

$$F_T(t) = t^{\lambda}, 0 \leq t \leq 1,$$

where

$$\lambda = \kappa\delta.$$

Second, we suppose that the discrete random variable

$$N$$

follows the geometric distribution with probability generating function

$$P_N(z) = \frac{pz}{1 - qz}, 0 < p < 1, q = 1 - p \tag{10}$$

and the discrete random variable

$$S$$

follows the geometric distribution with probability generating function

$$P_S(z) = \frac{\theta z}{1 - \pi z}, 0 < \theta < 1, \pi = 1 - \theta. \tag{11}$$

From (4), (10) and (11) it easily follows that the random sum

$$R = S_1 + S_2 + \ldots + S_N$$

follows the geometric distribution with probability generating function

$$P_R(z) = \frac{cz}{1 - \ell z}, \tag{12}$$

where

$$c = p\theta$$

and

$$\ell = 1 - p\theta.$$

Moreover, we suppose that the nonnegative random variable

$$X$$

follows the exponential distribution with distribution function

$$F_X(x) = 1 - e^{-\mu x}, x \geq 0, \mu > 0. \tag{13}$$

From (5), (12) and (13) it easily follows that the distribution function of the stochastic model is given by

$$F_T(t) = \frac{1 - e^{-\mu t}}{1 - \ell e^{-\mu t}}, \ t > 0.$$

It is readily understood that the theoretical results of the present section substantially facilitates the applicability in various practical disciplines of the stochastic model formulated by the previous section.

4 Applications in Systems and Processes

The present section of the chapter concentrates on the establishment of applications of the formulated stochastic model. It is shown that the model can be useful for managing risks threatening intelligent multimedia systems and investigating the performance of educational processes.

During the last two decades, the coordinated and secure storage, processing, transmission, and retrieval of multiple forms of information such as audio, image, video, animation, graphics, and text have been transformed into a dynamic field of research in modern informatics. It is generally accepted that the term multimedia systems is suitable for this field of research [8–10]. More precisely, this term is used to mean systems that accomplishes distinct tasks of manipulating pieces of information, mainly of different forms, but these pieces of information are readily recognized as a totality. Nowdays, multimedia systems have been altered to very usual tools for a very wide variety of activities. It is taken for granted that interactivity, personalization and adaptivity must constitute fundamental elements of intelligence multimedia systems [9, 11]. Such systems have to be interactive not only through ordinary modes of interaction. In addition, they have to support alternative modes of interaction, such as visual or lingual computer-user interfaces, which depict them more captivating, user friendlier, more approachable and more instructive. It is readily understood that an essential purpose of intelligent multimedia systems is their capacity for dynamic adaption to their users [9, 12]. For this reason, intelligent multimedia systems have to expand at all stages of processing for the implementation of their essential and minor purposes. Such an implementation incorporates additional research and advancement of low-level data processing for security, compression, transmission, clustering, classification, and retrieval of information [9, 10]. Research of this kind results to the advancement of novel and very capable intermediate level intelligent multimedia systems suitable for supporting systems for rights management and licencing, environmental information systems, home and office automation, human and object monitoring, and information tracking [8–10]. Such intermediate-level intelligent multimedia systems constitute the structural elements of high-level intelligent multimedia services which are very useful for digital libraries, e-learning, e-government, e-commerce, e-entertainment, e-health, e-legal, and other activities of particular practical importance [9, 13–16]. Services provided by intelligent multimedia systems have recently made a very impressive advancement, as they are incorporated in advertisement, creative industry, entertainment, art, education, engineering, medicine, mathematics, scientific research, and other theoretical and practical activities. The expansion of intelligent multimedia services is extremely fast, as technological advancement is directed to the strong support of modern complex organizations. Below, it is shown that the implementation of some essential and minor purposes of intelligent multimedia systems can be supported by the formulated stochastic model.

We suppose that the discrete random variable

$$N$$

denotes the number of different categories of risks threatening an intelligent multimedia system at some given time point in the future. We call this time point 0. We also suppose that the discrete random variable

$$S_n, n = 1, 2, \ldots$$

denotes the number of risks of the nth category. It is easily seen that the random sum

$$R = S_1 + S_2 + \ldots + S_N$$

denotes the number of risks contained in

$$N$$

such categories. Moreover, we suppose that the nonnegative random variable

$$X_r, r = 1, 2, \ldots$$

denotes the occurrence time of the rth risk contained in some of the

$$N$$

categories of risks. It is readily understood that the stochastic model

$$T = \min\{X_1, X_2, \ldots, X_R\}$$

denotes the smallest risk occurrence time. If the discrete random variable

$$N$$

denotes the number of different categories of catastrophic risks threatening an intelligent multimedia system at the time point 0, then it is quite obvious that the time interval

$$[0, T)$$

is of particular theoretical and practical importance for making decision concerning the performance and evolution of this system under the presence of such categories of risks. More precisely, such decisions can be valuable for selecting, analyzing, assessing and implementing combinations of risk control operations for optimal proactive treatment of the categories of catastrophic risks threatening the performance and evolution of an intelligent multimedia system.

It is quite obvious that the proposed application of the formulated stochastic model can be of some particular theoretical and practical importance for a very wide variety of systems. In particular, the general structure of information systems makes quite clear that the formulated stochastic model can substantially support the extremely important role, in theory and practice, of these systems. The general recognition of intelligent multimedia systems, as a leading kind of information systems, makes necessary a comment on the role of the formulated stochastic model for investigating and handling risks threatening this kind of information systems.

From the fact that risks threatening intelligent multimedia systems can be categorized by many different ways it easily follows that the formulated stochastic model can be suitable for managing such risks. Two fundamental ways of forming categories for risks threatening intelligent multimedia systems are the following.

First, we suppose that the research activities incorporating an intelligent multimedia system are categorized according to their purpose. Hence, at the time point 0, a random number of categories of research activities incorporates this system and each category contains a random number of research activities. From the fact that research activities are frequently recognized as causes of risks threatening an intelligent multimedia system, then it is readily understood that the formulated stochastic model is suitable for investigating the behavior of such a system under the presence of a random number of competing research activities forming a random number of different categories.

Second, we suppose that the persons making use of an intelligent multimedia system are categorized according to their profession. Hence, at the time point 0, a random number of categories of persons corresponds to this system and each category contains a random number of persons. Since human errors are very common causes of risks threatening an intelligent multimedia system, then it is easily seen that the formulated stochastic model is suitable for investigating the behaviour of such a system under the presence of a random number of competing human errors forming a random number of different categories.

The significant contribution of intelligent multimedia systems to many areas of education constitutes a very good reason for considering the suitability of the formulated stochastic model for assessing educational processes [6, 17, 18]. We suppose that the discrete random variable

$$N$$

denotes the number of different categories of persons having attended the same educational process. We also suppose that the discrete random variable

$$S_n, n = 1, 2, \ldots$$

denotes the number of persons of the nth category. Hence the random sum

$$R = S_1 + S_2 + \cdots + S_N$$

denotes the number of persons contained in

$$N$$

such groups. Moreover, we suppose that each person starts the completion of a given task at the time point 0 and the nonnegative random variable

$$X_r, r = 1, 2, \ldots$$

denotes the time required by the rth person, contained in some of the

$$N$$

categories, for the completion of the given task. Hence the stochastic model

$$T = \min\{X_1, X_2, \ldots, X_R\}$$

denotes the smallest completion time for the given task. It is obvious that the fundamental factors and the mathematical structure of the formulated stochastic model can be useful for assessing the educational process. More precisely, this model can provide analysts with valuable probabilistic information for investigating the performance of the educational process.

5 Conclusions

The use of a random sum of discrete random variables for formulating a stochastic model, being the minimum of a random number of nonnegative random variables, constitutes the theoretical contribution of the chapter. Moreover, the practical contribution of the chapter consists of providing applications of the formulated stochastic model in managing risks threatening intelligent multimedia systems and assessing educational processes.

References

1. Leveson, N., Harvey, P.: Analyzing software safety. IEEE Transform. Softw. Eng. **9**(5), 569–579 (1983)
2. Littlewood, B., Strigini, L.: The risk of software. Sci. Am. **267**(5), 38–43 (1992)
3. Parker, D.: Fighting Computer Crime. Wiley, New York (1998)
4. Schweitzer, J.: Protecting business information. Butterworth-Heinemann, Boston (1996)
5. Wright, M.: Third generation risk management practices. Comput. Fraud Secur. **1999**(2), 9–12 (1999)
6. Artikis, C.T., Artikis, P.T.: Learning processes and information management operations utilizing a class of random sums. Collnet J. Scientometrics Inf. Manage. **3**, 31–38 (2009)

7. Artikis, P.T., Artikis, C.T.: Recovery time of a complex system in risk and crisis management operations. Int. J. Decis. Sci. Risk Manage. **1**, 104–111 (2009)
8. Jordan, K.: Multimedia : From Wagner to Virtual Reality. Norton, New York (2001)
9. Tsihrintzis, G., Jain, L.: Multimedia services in intelligent environments : an introduction. In: Tsihrintzis, G., Jain, C. (eds.) Studies in Computational Intelligence, vol. 120, pp. 1–8. Springer-Verlag, Heidelberg (2008)
10. Vaughan, T.: Multimedia: Making it Work, 6th edn. McGraw Hill, Burr Ridge (2004)
11. Mandal, M.: Multimedia Signals and Systems. Kluwer, Dordecht (2002)
12. Rao, K., Bojkovic, Z., Milovanovic, D.: Multimedia Communication Systems: Techniques, Standards, and Networks. Prentice-Hall, Englewood Cliffs (2002)
13. Bhatnagar, S.: e-Government : From Vision to Implementation—A Practical Guide with Case Studies. Sage Publications, Thousand Oaks (2004)
14. Clark, R., Mayer, R.: e-Learning and the Science of Instruction : Proven guidelines for Consumers and Designers of Multimedia Learning. Wiley, New York (2008)
15. Prabhakaran, B.: Multimedia Database Management Systems. Kluwer, Dordecht (1997)
16. Witten, I., Bainbridge, D.: How to Build a Digital Library. Morgan Kaufmann, Los Altos (2003)
17. Constandanche, G.: Models of reality and reality of model. Kybernetes **29**, 1069–1077 (2000)
18. Eriksson, D.: A framework for the constitution of modelling processes : a proposition. Eur. J. Oper. Res. **145**, 202–215 (2003)

Author Biographies

Constantinos T. Artikis Constantinos T. Artikis is a risk analyst at Ethniki Asfalistiki, Insurance Company. His research interests concentrate on the mathematical modelling of risk management operations and information management. His published work has appeared in journals such as *Journal of Discrete Mathematical Sciences and Cryptography, Journal of Statistics and Management Systems, Journal of Information and Optimisation Sciences* and *International Journal of Applied Systemic Studies and Kybernetes.*

Panagiotis T. Artikis Panagiotis T. Artikis is an Adjunct Lecturer at the University of Athens. His research interests include stochastic models, enterprise risk management and sociological theories of risk. He has published in journals such as *International Review of Economics and Business, Computers and Mathematics with Applications, Journal of Statistics and Management Systems* and *Journal of Information and Optimisation Sciences and Kybernetes.*

Printed in the United States
By Bookmasters